DR. PAUL MAJETT ED.D.

AI Revolution

A Turning Point in Education

Contents

Preface

I decided to write this book while reflecting on 25 years of higher education. As a Business Consultant, I've developed strategies for Leaders, Entrepreneurs, and Professionals that focus on successfully merging and acquiring businesses, adapting to rapidly evolving environments, and developing effective organizational frameworks for their workforces. My pedagogical approach emphasizes student-centered curricula, where each student is the driver of their learning journey.

My professional experience spans over 25 years, and I've worked in more than a thousand workplace environments. I began volunteering at 11 and entered the workforce at 14. I have worked in various industries, including Restaurant, Retail, Manufacturing, Medical, Financial, Legal, Real estate, Insurance, and Higher education, across more than a dozen Fortune 100 companies.

I am passionate about teaching and enjoy sharing my knowledge and experiences with the world. It is essential to foster environments where people feel empowered to make their own decisions, build self-confidence, and expand their critical thinking skills. I study history, enjoy real-time, and I am passionate about the future.

I encourage all readers to make a conscious effort to consume and digest each word with intention, so that they may experience the revelations that propel them to succeed in life. I have always had an intrinsic motivation to help others achieve their goals. I hope that this book will empower human resilience and adaptability in the face of a rapidly changing world.

Introduction

This century has witnessed a significant shift in educational paradigms, with a growing emphasis on adult education, higher education, and skilled trades. The accessibility of student loans, which enables the broader proliferation of these industry-aligned programs, contributed to this development. Universities, recognizing the shift, have forged stronger partnerships with industry leaders.

Curriculum development committees, often including representatives from specific industries, align course content with immediate and projected workforce needs. It has led to the introduction of specialized degrees, certificate programs, and apprenticeships designed to equip students with tangible, in-demand skills.

The world is witnessing rapid industry evolution, accompanied by growing demand for a range of proficiencies, including soft skills. As artificial intelligence continues to weave itself into the fabric of daily life, its influence on the job market has become undeniable. While AI brings about unprecedented efficiency and convenience, its pervasive presence has also created a unique challenge: a deficit in human skills.

This new era reveals a skill gap that threatens the harmony between man and machine. As individuals, we rely heavily on AI for tasks that require analytical thinking and technical prowess, but this very reliance has created a void. Soft skills —such as effective communication, creative problem-solving, and emotional intelligence —are being overlooked, creating a deficiency that could hinder progress.

Another intriguing paradox that has emerged is the new reality: while technology connects people globally, it has also isolated them. The more individuals depend on AI for interaction, the wider the gap grows, leaving a void that human connection and creativity could only fill. Educators and employers race to address this unique challenge. They scramble to equip individuals with the soft skills they need to succeed in a changing world.

I

Part 1: American Dream: The Pursuit of a Better Life

Part 1 is an analysis set within the contemporary American educational and economic landscape, tracing developments from the late 20th century to the present day. It addresses a relevant and timely topic: the intersection of education, student loans, and AI. This section of the book provides a historical overview of the Department of Education and the evolution of education in the 21st century.

1

Adapting to the Changing Landscape of American Education

21st Century Education in the U.S.

The Department of Education, established as a cabinet-level agency in 1980, has played a pivotal role in shaping the United States' educational landscape. Its origins, however, extend further back in history. The department's earliest predecessor was the Office of Education, founded in 1867 as part of the Department of the Interior. This office collected educational data and disseminated information, reflecting the federal government's growing interest in education.

Over time, the office evolved, and in 1953, it became the Department of Health, Education, and Welfare, reflecting the growing importance of education on the nation's agenda. The department has since undergone several structural changes, reflecting the developing priorities and needs of the American education system. The modern department was established as a cabinet-level agency in 1979, following Jimmy Carter's campaign to separate the agency from what is now the Department of Health and Human Services. The Department of Education received approximately $229 billion of the $9.7 trillion allocated to federal agencies in 2024.

The Department of Education's Office for Civil Rights (OCR) is responsible for ensuring that students do not experience discrimination. It fields complaints, oversees compliance reviews, and works with schools to develop and implement corrective actions to bring them into compliance. Congress allocated $140 million to the OCR in 2024, but ultimately, it is the White House that decides what the OCR prioritizes.

One of its key roles has been to ensure equal access to quality education for all Americans, regardless of their background or location. Title 9 is a high-profile federal law that the Department of Education is responsible for enforcing. It prohibits discrimination based on sex in any education program or activity that receives federal financial assistance. This mission has led to several initiatives and policies. These include expanding federal financial aid and promoting education and technology. Both have affected education in the 21st century.

The department is also responsible for issuing Pell Grants, which are typically awarded to undergraduate students based on their financial need and do not usually require repayment. It also manages Title I funds, a federal aid program for K-12 schools serving low-income students. This program provides funding to students living in areas of concentrated poverty.

A commitment to inclusivity and a forward-thinking approach has characterized the Department of Education's journey through the 21st century. Over time, the department recognized the need to adapt to a transforming world. Society is developing, and with it are the demands of the American people. The department has embraced the challenge, understanding that education is the cornerstone of progress. In the early years of the new millennium, a shift occurred toward greater emphasis on adult education and skilled trades. The shift marked a significant step towards accessibility and equality.

Recognizing that education is a lifelong pursuit, the department developed initiatives to encourage and support adults returning to education. Whether it was upskilling, retraining, or pursuing new passions, the department fostered a culture of lifelong learning. Additionally, it highlighted the skilled trades as a means to address the need for a talented workforce to meet the growing economic demands. This era witnessed a celebration of vocational paths, offering alternatives to traditional academic routes and highlighting the value of all careers.

The evolution of education in the United States in the 21st century has been a dynamic journey, with a focus on various sectors of learning. Adult education, higher education, and skilled trades have undergone significant transformations, adapting to the needs of a changing society and technological advancements. The burgeoning demand for skilled labor in sectors such as advanced manufacturing, renewable energy, and cybersecurity has influenced the reconfiguration of higher education and vocational training programs.

Predictable Progression

The initial alignment of education programs with industry demand, a hallmark of the early 21st century, operated on a model of predictable progression. Universities and vocational institutions analyzed workforce projections and collaborated with industry leaders to design curricula for sectors such as advanced manufacturing, renewable energy, and cybersecurity. Their efforts involved creating specialized degrees and certificate programs, often with direct input from companies seeking graduates with specific skill sets.

Student loans helped people access these programs, enabling a wider range of individuals to acquire tangible, in-demand skills, which led to positive changes in employment statistics and salary potential. The alignment between educational offerings and industry demand has not always conformed to the traditional model of a four-year institution. Community colleges and technical schools existed, adapting their vocational training to meet the developing requirements of the job market.

Programs in areas such as advanced welding, mechanical, and data analytics saw significant investment and expansion. As learning lifelong gained popularity, Universities and private organizations created adaptable educational paths. These paths enabled adults to enhance their skills and learn new ones without pursuing a full degree.

Employment statistics showed the impact of this alignment, with graduates reporting lower unemployment rates and higher starting salaries. The intentional alignment between educational offerings and industry demand that characterized the early 21st century shifted notably in the subsequent years. The emphasis on specialized degrees and certificate programs, intended for direct applicability, revealed limitations. Industries began to pivot, often requiring adaptability and cross-disciplinary knowledge that traditional, focused programs struggled to provide. Lifelong learning remains meaningful, but it has become increasingly complex. Individuals not in structured educational settings must update and re-evaluate their skills.

AI-driven Innovation

The rapid pace of technological advancement, coupled with evolving global market dynamics, has created a more fluid and unpredictable landscape. The emergence of artificial intelligence (AI) has also disrupted the established alignment. AI's capacity for automation, data analysis, and even creative problem-solving has rendered some specialized skills that education programs once considered obsolete or devalued.

The once-robust partnership between academia and industry has weakened as the pace of AI-driven innovation outpaces educational institutions' ability to update curricula in real time. The robust alliances among universities, community colleges, and industry leaders have eroded. Curricula crafted to meet specific workforce needs are now lagging as the pace of innovation has accelerated.

A significant disconnect exists because AI-powered tools and platforms have rendered skills that were once in high demand obsolete. The pathways to acquiring tangible, in-demand skills have become obscured, leading to growing concern about graduate employability in sectors affected by AI adoption. The misalignment has introduced new challenges for both students and the financial structures supporting their education. The path to gaining valuable skills has become unclear. There is a misunderstanding between what schools teach and what jobs require.

Amidst a rapidly evolving landscape, a new generation of learners is emerging, and the burgeoning influence of AI and technology is shaping their educational journeys. These students, unlike their predecessors, who sought specialized vocational skills, navigate a world in which the very nature of work has undergone significant change. They are the digital natives, the ones who understand the power of algorithms and the potential of machine learning. They no longer confine their learning experiences to traditional classrooms.

The transformation, while offering unprecedented flexibility and efficiency, also introduces an additional layer of complexity to the student loan narrative. These loans help learners invest in their ability to adapt and learn with AI, not just to gain skills. Educational institutions are grappling with how to foster these qualities within AI-infused curricula and how to assess them. The challenge for these students lay in translating this augmented learning into tangible career advancement. While AI offers powerful tools for analysis and problem-solving, the "human" skills – critical thinking, creativity, emotional intelligence, and ethical reasoning – have become prized.

Traditional loan-supported educational pathways have struggled to provide the agility needed for continuous upskilling and reskilling amid AI's transformative impact. These pathways have made the world more complex and uncertain. The promise of a better life through education is now contingent on navigating AI-driven change.

Individuals leveraging student loans to gain skills in fields now dominated or augmented by AI find that their investment does not guarantee the expected direct return. Focused programs, supported by student loans, often failed to prioritize the adaptability and cross-disciplinary knowledge needed for AI integration. Student loans are still accessible. However, they are now more focused on programs. These programs offer foundational expertise but do not guarantee immediate job placement in high-demand fields. This divergence has been a growing concern for graduate employability in specific sectors.

Some graduates possess skills that exceed current demand, while newer technologies and methodologies have superseded them; however, some individuals continue to use their education to advance their careers. Previously, people considered student loans as investments with clear returns. Now, they have become a burden for some. Their expertise does not always pay off in the current economy. This changing educational environment, shaped by progress and societal shifts, poses new challenges. Academic institutions and individuals face these challenges as they seek knowledge and a better life.

Implications of Student Loan Debt and the Challenges Posed By AI

As we trace this journey, it is imperative to understand the role that student loans have played in shaping the educational landscape. Student loans have played a critical role in making these shorter, more focused educational opportunities accessible, allowing individuals to adapt their skill sets to the dynamic economic landscape. The history of student loans is a complex and intriguing narrative that intertwines with the very foundation of the American dream: the pursuit of knowledge and a better life.

Adult education, higher education, and skilled trades have undergone significant transformations, adapting to the evolving needs of a changing society and the inevitable progression of technological advancements. For teenagers entering this changing landscape, the path forward often involves financial planning from an early stage.

The decision to pursue higher education has, for many, become linked to the process of navigating loan applications, with parents and guidance counselors often playing advisory roles. High school students, even before graduation, engage with information about college applications and the associated costs. These decisions include attending financial aid workshops, researching scholarship opportunities, and understanding the implications of taking out student loans.

The story of student loans has become both an opportunity and a burden for many. Student loans have been a pivotal aspect of higher education in the US for decades. As the cost of tuition and living expenses has soared, loans have become a necessary tool for many students to pursue their academic goals. Debt remains a burden for many. However, those who use loans to learn skills in high-demand fields view them as a worthwhile investment. Industry needs and financial aid support drove this shift in the education sector. This shift enabled many Americans to achieve a better life through learning and career growth.

For some, loans provide a pathway to prestigious universities and a chance at a brighter future. Yet, for others, it has become a weighty chain, binding them to a lifetime of debt and financial struggle. This dual nature of student loans has had a profound impact on the lives of countless individuals and influenced the very structure of American society.

The impact of student loans extends beyond the immediate decision to attend college. As many young adults transition into higher education, the financial mechanisms that shape their daily lives include grants, scholarships, and loans that cover expenses such as textbooks, tuition, and living costs. For many teenagers, the administrative processes surrounding student loans have become an integral part of the higher education experience.

The narrative of student loans solidifies as a strategic investment for many, promising a clear return through career advancement in defined fields. The accessibility of student loans remains a critical factor, but the focus is shifting. People view loans as enabling participation in dynamic, project-based learning environments that mirror real-world problem-solving, rather than funding a degree or certificate. However, it increases the risk.

Student Loan Debt (ROI)

The developing educational landscape, characterized by personalized learning and lifelong adaptation, further complicates the student loan equation. The ROI of these academic pursuits is changing. It's less about a specific job and more about developing intellectual skills for an AI-driven future.

This issue makes debt a heavier burden if the skills don't lead to economic success. The very definition of "career advancement" has morphed for these students. A more fluid, iterative journey has replaced the singular, linear career path of previous generations. In this context, student loans have become inextricably linked to the concept of continuous adaptation.

Student loans are not just for initial education anymore. They have had a ripple effect on consumer spending and economic growth. Financial institutions and government policy are under pressure due to the sheer volume of outstanding student loan debt. Interest in the debate over loan forgiveness, rates, and financing has grown. Although intended to help, the original system is now causing financial problems for many. This widespread financial strain fueled a national discourse centered on the efficacy and sustainability of the existing student loan system.

The US Department of Education operates in many ways, much like a bank. One of the department's most significant responsibilities is disbursing federal student loans and managing loan-forgiveness programs. Policy discussions have proliferated, encompassing proposals for loan-forgiveness programs, reforms to interest-rate structures, and the exploration of alternative financing models for higher education and skilled trades, such as micro-scholarships or income-share programs. This widespread financial strain has catalyzed a national dialogue regarding the efficacy and long-term sustainability of the prevailing student loan system.

The realization that the implications of this debt crisis extend beyond individual financial well-being has heightened the urgency of these discussions and affected the nation's overall economic health. The student loan debt crisis, which affects not only the borrower, has altered the financial trajectories of American families. US education costs, especially at universities, have grown faster than inflation for many years. Student loans have both eased and worsened this problem. Access to student loans is vital for individuals undergoing AI-driven changes.

The debt's weight, along with job market uncertainty, necessitates a reevaluation of our approach. The weight includes education financing and its impact on individual finances and the American dream. Other developed nations offer more affordable tuition or better public funding for higher education. This results in less student debt for their citizens.

Education in Other Developed Nations

Countries like Germany, for instance, provide tuition-free university education to both domestic and international students, thereby eliminating the substantial financial burden that characterizes the American experience.

This contrast highlights a significant difference in how they enable access to knowledge and career advancement, as well as the economic implications for individuals and society.

The student loan debt crisis in the U.S. is a systemic issue. It presents a challenge to social mobility, economic opportunity, and the American dream. The need for continuous adaptation and lifelong learning, while crucial in the current economic climate, becomes an even more significant undertaking when compounded by the enduring burden of educational debt.

Comparative financial models from other developed nations, where public funding or lower tuition costs result in lower student debt, highlight the unique burden placed on American citizens. The AI-augmented job market requires continuous adaptation and lifelong learning. Educational debt makes this more complicated. The American Dream and social mobility are affected across generations.

The Grand Default & Debt Forgiveness

The escalating student loan debt in the United States presents a multifaceted challenge, affecting credit scores and long-term financial planning. Unlike many other forms of unsecured debt, federal student loans possess a unique characteristic: discharge is rare but possible under the undue hardship tests. Therefore, people must still repay these loans even during severe financial distress or personal hardship, unlike medical debt or personal loans, which the borrower cannot discharge in bankruptcy.

Certain loan forgiveness programs exist for specific situations. However, people still debate whether to forgive all student loan debt. Many borrowers expect to be repaying their loans for a long time. Due to the permanent nature of student debt, its impact on credit reports can last for many years, influencing a borrower's chances of securing future loans, mortgages, or some job prospects.

Defaulting on student loans has impacts that extend beyond financial consequences. It often leads to a downward spiral of consequences, including damaged credit scores, difficulty in securing future loans, and even legal repercussions. The impact of student loan debt on an individual's credit profile is substantial and can manifest in several ways. Even though they are beneficial, missed or late payments can overshadow consistent on-time payments. Failing to pay federal student loans can damage your credit score. Defaults can make it hard to rent an apartment, get a car loan, or secure reasonable interest rates.

The extended repayment periods often associated with significant student loan balances can mean that this debt occupies a substantial portion of a borrower's credit utilization ratio for an extended period. The dilemma also limits the credit available for other financial needs and investments. Poor management's repercussions are harsh and enduring because it's difficult to discharge or forgive. These repercussions have a long-term impact on a person's financial situation, affecting their ability to participate in economic activities.

The Great Recession of 2008 further complicated the narrative. As the economy faltered, many found themselves unable to find employment, and the burden of student loans became even more pronounced. Over the years, variable interest rates and shifting economic conditions have affected the repayment journeys of many borrowers. Some graduates have been burdened by rising interest rates, making their monthly payments a significant strain on their finances. This period witnessed a surge in default rates, with far-reaching consequences for both individuals and the lending institutions.

The impact of this crisis would shape policy discussions and influence approaches to addressing the growing student debt crisis. It catalyzed the role and accessibility of higher education in the country, as borrowers struggled to keep up with the compounding interest on their debts. The consequences of default have extended far beyond financial institutions; they have impacted the lives and prospects of individuals, becoming a significant hindrance in their pursuit of the American dream.

The Color of Debt

Minority communities, including Black and Hispanic populations, have been affected by student loan debt in the United States. Historical and ongoing racial disparities in wealth accumulation, income, and access to financial resources have contributed to these communities bearing a heavier burden of student loan debt. Discriminatory lending practices, unequal access to quality K-12 education, and systemic barriers to higher-paying employment often cause this disparity. These factors existed before student loans were available, but student loans exacerbate them. Individuals from these communities may need to borrow more to finance their education, leading to higher initial debt loads.

The difficulty of discharging student loan debt through bankruptcy exacerbates existing disparities, as affected communities often lack financial alternatives when facing repayment issues. The long-term consequences of this disproportionate debt burden manifest in several observable ways. Graduates from Black and Hispanic communities, often carrying higher student loan balances, experience similar deferrals of major life milestones such as homeownership and family formation, but with amplified severity. This financial precariousness can limit intergenerational wealth transfer, perpetuating cycles of disadvantage.

The impact on credit scores, while universal, can be detrimental to minority borrowers, who may have fewer opportunities to rebuild their credit because of systemic economic inequalities. Adaptation and lifelong learning are essential, but are more challenging for those with educational debt. This dilemma affects social mobility and the American dream.

The systemic nature of student loan debt, therefore, extends beyond individual financial management to societal implications. Deferred life milestones, such as homeownership and family formation, impact economic sectors that rely on consumer spending and investment. This intricate web of challenges influenced the decisions and life paths of those affected, limiting their opportunities for homeownership, starting businesses, or pursuing specific careers.

Loan defaults often cause increased stress, anxiety, and a sense of diminished hope for the future, and the toll on people's emotions and psychology is significant. The simultaneous occurrence of all these problems causes many people to stop repaying their loans, as rising interest rates make it too expensive. The impact of these defaults is profound, not just on financial institutions but also on borrowers' lives and prospects.

The legal repercussions add further strain, with some facing wage garnishments or even bankruptcy. This intricate web of financial and legal challenges shapes borrowers' life paths, leading them to make decisions that might otherwise have seemed unthinkable. The student loan system, intended to democratize access to education, has, through its structure and the existing socio-economic landscape, become a mechanism that can reinforce and even widen existing racial and economic inequalities.

Compared to the educational financing models of other developed nations, racial disparities are more pronounced. These models often lead to less student debt and improved social mobility. Thus, the United States needs to solve the student loan debt crisis. It must also address the systemic issues that affect minority borrowers. The solution will require a re-evaluation of how people access education and career opportunities.

AI's Impact and the Challenges It Poses to Traditional Education Models

The pervasive integration of artificial intelligence (AI) across sectors of the economy has introduced new dimensions to both educational attainment and the management of student loan debt. AI-powered adaptive learning platforms and personalized educational pathways, while offering increased flexibility and efficiency, also strengthen the educational financing landscape.

Students use AI tools for research, data analysis, and even to enhance their learning processes, aiming to gain skills that remain relevant in an AI-augmented job market. However, positing these skills, often facilitated by student loans, does not always guarantee immediate economic returns, especially as AI continues to automate tasks requiring human expertise. This progression creates an ongoing cycle of upskilling and reskilling, which can lead to increased borrowing and debt, despite the availability of AI tools that could aid financial management and repayment.

The intersection of student loan debt and AI presents a complex challenge for individuals and policymakers alike. AI technologies have developed to assist in financial planning, debt management, and even to identify potential career pathways with a higher likelihood of loan repayment success. These tools can analyze spending habits, predict future earning potentials based on skill acquisition, and offer personalized strategies for managing student loan obligations.

AI solutions are accessible, but their effectiveness differs. Additionally, they fail to address the rising cost of education and the resulting debt crisis. We need to criticize the systemic nature of student loan debt and its impact on minority communities. We must also ensure fair deployment of AI financial solutions to avoid worsening disparities in access to knowledge and opportunities.

Technological advancements are acting as a primary catalyst for these multifaceted shifts. The pervasive influence of the internet and digital tools facilitates greater access to information, fostering environments that support self-directed learning and enable the creation of innovative educational content.

Virtual Classrooms and Learning Management Systems

Online courses, virtual classrooms, and sophisticated learning management systems have transformed the way people disseminate and acquire knowledge in higher education and adult learning. These advancements have created new demands for specialized skills, affecting the curriculum and delivery methods in skilled trades, as well as the use of virtual reality simulations for training in complex machinery operations in the future.

Additionally, augmented reality will help with diagnostics and repairs. These changes represent a shift away from traditional hands-on methods. Technological advancements have played a pivotal role in this transformation, offering new avenues for learning and knowledge dissemination.

Online platforms in higher education offer more opportunities due to increased accessibility. However, this also raises concerns about fair resource distribution and the potential for a growing digital divide. In the realm of adult education, the rapid obsolescence of specific skill sets causes continuous program adaptation to maintain relevance. Skilled trades now require more than just manual dexterity.

Candidates need digital interface and data analysis skills. This necessity reshapes the ideal candidate and demands updated training. The synergistic interplay between these diverse educational sectors and the relentless march of technological progress has generated a complex and dynamic educational landscape.

21st Century American Education Framework

The 21st century has witnessed a palpable recalibration of the American educational framework. A significant restructuring of the American educational system has marked it, with a pronounced emphasis on adaptation and diversification. Adult education has experienced a resurgence, driven by the imperative for lifelong learning amid rapid industrial and technological shifts. Upskilling and reskilling programs have become commonplace for individuals seeking to adapt to emerging job markets or transition into new careers.

Today, many universities are adapting to the digital revolution, integrating online learning platforms and hybrid models to serve a more diverse, dispersed student body. This adaptation has sparked a renewed emphasis on skilled trades. As automation advances, specialized technical skills are in high demand. Fields such as advanced manufacturing, renewable energy, and cybersecurity require specialized expertise and knowledge. This requirement has led to renewed interest in vocational training and apprenticeships. These are valuable paths to economic stability.

Adult education, once a niche sector, has experienced a notable resurgence as the imperative for lifelong learning has become undeniable amid swift industrial and technological transformations. The great promise of the American Dream, with its pursuit of knowledge and opportunity, has appeared to falter for many, the very students who took out loans to improve their prospects find themselves in a Catch-22. Students question whether their degrees were valuable.

The dream of a better life through education seems unattainable for many. As higher education becomes more expensive, so do the skills needed to secure a well-paying job. Students feel pressured to take on even more debt to keep up with the demands of a changing job market. The skilled trades, once a viable alternative to a traditional college degree, are not immune to these shifts.

As technology advances, the cost of equipment and training has risen, creating another barrier for those seeking to enter these fields. The skills they have gained, whether through adult education programs or prestigious universities, are not enough to secure stable employment.

The job market continues to leave many graduates struggling to keep up. Graduates accumulate debt and lack the specific skills the ever-changing economy demands. This skills gap continues to widen as technological advancements outpace educational institutions' ability to adapt their curricula.

As the 21st century progresses, the cost of education continues to rise, outpacing inflation and wage growth. As the cost of living continues to increase in the United States, the idea of taking on more debt to acquire additional skills has become increasingly unaffordable for many. Student loan burdens have created a cycle of disadvantages. Those with debt feel trapped and struggle to develop the skills they need for the job market.

Historically, as more students pursue higher education, the job market has become saturated. Employers can now be more selective, often requiring advanced degrees and specific skill sets. Their pickiness has created a vicious cycle where students take on more debt to gain a competitive edge, only to find themselves in a crowded market with limited opportunities.

The very institutions that empowered individuals were now shackling them with financial chains. Those in this predicament have not missed the irony: the pursuit of knowledge has become a trap. For many, their loans remind them of the opportunity they grasped at a tremendous cost. The American Dream, with its promise of upward mobility, is becoming increasingly elusive for those burdened by student debt. As the century progresses, the very loans that empower are shackling.

The job market, once a fertile ground for aspiring graduates, has become a barren landscape, with automation and economic shifts reducing the number of available positions. Those who invest in their education find themselves in a cruel paradox: equipped with knowledge but lacking the practical skills demanded by a rapidly changing job market.

This skills gap leaves many graduates struggling to find employment, with their degrees becoming increasingly irrelevant in a world that prioritizes technical proficiency over academic achievement. The traditional path of higher education is no longer a guaranteed route to success. Against this backdrop, the Department of Education has navigated a path of evolution, adapting to the changing landscape of American education.

2

The Internet and AI's Impact on the American Classroom

Distance Learning's Journey

The historical trajectory of distance learning provides a crucial backdrop to understanding the internet's current transformative role. Educational institutions delivered distance learning through correspondence courses via postal mail. These ancient methods enabled dispersed individuals to access education, but they had limited interactivity and a one-way flow of information. These early iterations, though limited in interactivity, laid the groundwork for a more accessible educational model.

The subsequent advent of radio and television broadened the reach of distance learning, facilitating the broadcast of lectures and educational programming to a broader audience. These early iterations, though limited in interactivity, laid the groundwork for a more accessible educational model.

The Internet marked a fundamental shift in the paradigm. It transitioned education from a one-to-many broadcast model to a many-to-many net-worked interaction system. This evolution enabled both synchronous and asynchronous communication, allowing real-time discussions, collaborative document creation, and the establishment of virtual learning environments that mirror the dynamics of a physical classroom.

Additionally, the internet enabled both types of communication, including real-time discussions, collaborative documents, and virtual learning spaces. The internet's transformative effect on education spans all grade levels, extending the boundaries of the traditional classroom and fostering new modes of learning.

The Impact on K-12

For early elementary students, interactive online platforms have revolutionized the acquisition of foundational literacy and numeracy skills. These platforms have employed gamified lessons and deliver personalized feedback, enabling young learners to observe their peers' progress and navigate adaptive learning paths. This visual reinforcement of successful strategies fosters adoption of effective learning behaviors.

As students progress to middle school, the internet unlocks new avenues for collaborative learning. Virtual field trips expose them to diverse environments and cultures, while collaborative research projects facilitate global exchanges and foster a deeper understanding of them. During these interactions, students witness and emulate effective digital citizenship and problem-solving strategies demonstrated by classmates from diverse backgrounds, thereby cultivating essential 21st-century skills. High school students capitalize on the internet for more advanced academic pursuits.

With access to vast digital libraries, online tutoring services, and sophisticated virtual lab simulations, students can engage in an in-depth scientific inquiry and critical analysis. By observing how peers utilize these digital tools, students reinforce and refine their own sophisticated learning behaviors and research methodologies. For adult learners, online platforms deliver foundational literacy and numeracy skills through gamified lessons and personalized feedback. This delivery enables the observation of peers' successes and adaptive learning paths, reinforcing effective learning behaviors.

The Impact on Adult Education

At the university level and beyond, the internet facilitates access to vast digital libraries, open-source learning materials, and professional development webinars. Learners use interactive simulations and online communities to understand complex concepts. They observe how experienced people and peers use and discuss knowledge, which shapes their academic and professional behavior.

The continuous proliferation of specialized online learning platforms, coupled with the integration of artificial intelligence, has further refined these digital educational experiences. This proliferation has led to the creation of personalized learning journeys and expanded the scope of academic integration, transcending geographical limitations and fostering continuous intellectual growth.

Online information and diverse perspectives foster dynamic social learning. Students then adapt their research, communication, and critical thinking skills. These changes depend on observed effectiveness within a digital educational setting. The continuous flow of information and diverse perspectives available online encourages a dynamic process of social learning. Students adapt their research methodologies, communication styles, and critical thinking approaches based on the observed efficacy of these behaviors within a connected educational ecosystem.

The internet has significantly impacted the digital ecosystem, reshaping how people acquire, share, and apply knowledge throughout all stages of formal education. From kindergarten to postgraduate studies, the internet has not supplemented existing educational structures. Still, it has become an integral component, fostering an interconnected and strengthening learning environment driven by accessibility, collaboration, and intelligent adaptation.

The Pandemic's Impact

The COVID-19 pandemic served as a significant accelerator for integrating internet-powered learning, reshaping the educational landscape. Because physical schools closed everywhere, educational institutions had to switch quickly to online delivery models. These closures led to the swift adoption of digital tools and platforms across all grade levels, prompting educators and students alike to adapt to a new mode of instruction and learning.

The reliance on virtual classrooms, digital resources, and online collaboration tools intensified, highlighting both the potential and the challenges of internet-dependent education. This period propelled the adoption of technologies that might have otherwise taken years to become mainstream. It fostered innovation in online pedagogy, curriculum adaptation for digital environments, and the creation of strategies to maintain student engagement and well-being.

The pandemic showed the internet's ability to maintain education during crises. It also changed teaching methods, digital skills, and the classroom experience, making the internet essential to global education. This digital ecosystem, energized by the pandemic's necessity, has also given rise to new forms of student interaction and engagement, influenced by AI. Students are modeling and refining their cognitive and interactive behaviors within a globalized learning network. Their modeling is because of the pervasive nature of mediated social learning.

The success of one learned, amplified, and validated by intelligent systems, becomes a beacon for others, fostering a collective drive towards mastery. AI recognition and peer emulation create a feedback loop that reinforces effective strategies. This recognition transforms learning into dynamic, interconnected experiences. It transcends physical boundaries and uses collective intelligence.

Adopting Computer-Adaptive learning (CAL)

Computer-adaptive learning (CAL) systems, now integral to the AI-powered educational ecosystem, adjust the difficulty and presentation of material based on a student's real-time performance. These systems analyze response patterns, accuracy rates, and time spent on tasks to generate customized learning pathways. For instance, a student grappling with complex genetic sequencing might encounter more scaffolded explanations and simpler practice problems if her initial attempts prove challenging.

If they demonstrate rapid comprehension and accuracy, the system will introduce more advanced concepts and problem sets, thereby preventing disengagement due to boredom. This constant calibration ensures that each student operates within their optimal learning zone —a principle rooted in Vygotsky's zone of proximal development —facilitated by AI's analytical capabilities.

The observed efficacy of such adaptive interventions, as evidenced by students achieving learning objectives rapidly, reinforces the adoption of these CAL tools across the curriculum. The effectiveness of CAL integration with social learning amplifies its social learning mechanism. This observation-based learning, facilitated by AI's data analysis, allows students to witness the application of effective problem-solving techniques and then emulate them. Reinforcement comes from the AI's direct feedback and from observing peers' successes. The digital learning space highlights these successes, which encourage everyone to develop their skills.

This continuous feedback loop in computer-adaptive learning, along with peers' observable successes, nurtures a learning environment where students feel motivated to explore, experiment, and refine their approaches. Students gain knowledge from both the material and the effective digital methods shown by their peers. This overtime leads to the development of advanced research skills, critical thinking, and teamwork. Thus, the internet and AI, through CAL, have transformed education into a dynamic reinforced process, where collective learning and observed mastery intertwine with and accelerate individual progress.

Computer-adaptive learning (CAL) systems are now central to the adult education landscape, offering personalized and efficient pathways for lifelong learners. Professionals seeking to upskill or reskill find CAL platforms dynamically adjust content difficulty and the pace of instruction based on their existing knowledge and performance. For example, a marketing executive enrolled in an advanced digital analytics course might progress rapidly through modules on fundamental statistical concepts if their initial assessments show proficiency.

Conversely, the system would provide more detailed explanations and practice exercises for areas where their understanding is less firm, such as understanding the nuances of predictive modeling. This granular adaptation ensures that adult learners engage with material in their optimal learning zone, maximizing knowledge acquisition and minimizing wasted time.

The observed efficacy of these systems, demonstrated by improved certification pass rates and quicker skill acquisition, reinforces their widespread adoption in corporate training and professional development programs. Integrating CAL with social learning principles further enhances its impact in adult education. In collaborative learning environments, AI not only tailors individual learning trajectories but also analyzes group interactions and the successful strategies peers employ.

In a project management simulation, the AI could highlight the effective resource allocation methods of an anonymized participant. Other learners could then observe and emulate these methods. This observation-based learning allows professionals to witness practical applications of theoretical concepts and adapt their own methodologies.

AI feedback and observing colleagues' successful project outcomes reinforce these digital behaviors. This overtime leads to improved project management skills. CAL's continuous feedback loop, combined with peers' successes, motivates adult learners to experiment and refine their professional approaches. The AI tracks and highlights effective digital strategies. This process creates a pervasive digital reinforcement system, whether for data analysis or communication.

Professionals learn not only from the curriculum but also from the shown efficacy of their colleagues' digital methodologies. This learning experience develops advanced research skills, fosters critical thinking, and promotes collaborative problem-solving. Adult education becomes a dynamic process where individuals learn more effectively together, supported by social interaction and the observation of skill mastery.

Integrating computer-adaptive learning (CAL) systems within adult education aligns with established adult learning theories. those emphasizing andragogy and transformational learning. Knowles' theory, the basis of andragogy (Knowles, 1984), explains that adult learners direct their own learning, contribute substantial experience to the learning process, focus on problems, and possess internal motivation. CAL systems align with these principles by offering personalized learning paths that enable adult learners to leverage their prior knowledge and focus on specific areas of need.

The adaptive nature of these systems acknowledges the learner's experience by avoiding the need to revisit mastered content, respecting their self-direction. CAL facilitates a problem-centered approach by presenting real-world applications and challenges that adult learners can relate to their professional lives, fostering a deeper understanding and engagement.

Adult Learning Theories

Transformational Learning Theory, developed by Jack Mezirow, suggests that learning is transforming held assumptions and perspectives, often triggered by critical reflection and disorienting events (Mezirow, 2000). In CAL, exposure to diverse digital strategies and peers' observed successes can act as catalysts for such transformation. Others on the platform have shown how data-driven insights and effective methodologies can challenge an adult learner's assumptions.

For example, a marketing executive might encounter advanced predictive modeling. The system highlights successes and offers opportunities for emulation. This interaction enables critical self-reflection and may lead to a shift in understanding and application. It also aligns with the theory's emphasis on challenging existing frameworks and integrating new perspectives.

Emotional intelligence is key to this new era of learning, as students navigate their social and emotional landscapes. Goleman's model, with its five components, serves as a roadmap for students to understand and manage their own emotions, as well as those of others. Students practice self-regulation and self-awareness, enabling them to recognize and manage their feelings, and internal motivation encourages them to strive for improvement. Social skills and empathy would allow students to connect with and understand their peers, thereby fostering a collaborative and supportive environment. In this future classroom, students will develop soft skills, which combine interpersonal and people skills.

Goleman's descriptors of soft skills, including social awareness and relation-ship management, are explicit as students navigate their interactions with others. They learn to manage their emotions and develop self-awareness, recognizing their feelings and how they impact others. This emotional intelligence performance model guides students in identifying and utilizing their feelings, thereby shaping their behavior and interactions.

The Asia Society Center for Global Education (ASCGE) identifies key soft skills such as leadership, collaboration, and communication as essential for adaptability in adulthood. Interactive, immersive learning experiences facilitated by AI hone these skills. For example, AI platforms can create simulated work environments where students assume leadership roles, make decisions, and collaborate to solve complex problems. Here, they receive feedback on their emotional intelligence, encouraging self-awareness and self-management. Soft skills are integral to an individual's job performance and career development. By investing in these skills early on, students are better equipped to build social capital and succeed on their chosen paths.

AI and the internet have revolutionized the classroom, creating a dynamic and engaging learning environment that extends beyond traditional academic knowledge. As students graduate and enter the professional world, they do so with a well-rounded skill set, ready to adapt to the ever-changing demands of the future. Bandura proposed that people learn behavior through observation and understanding the consequences. This theory gains an additional dimension with the advent of AI.

Adult learners now have the unique opportunity to observe and interact with AI systems, learning not just from their peers and teachers but from intelligent machines. Social learning theories have provided a better understanding of how adult learners observe and emulate a diverse range of behaviors and skills from peers and experts worldwide. AI platforms facilitate connections between learners, creating global communities of practice.

For instance, an adult learner interested in programming can now connect with a worldwide community of programmers, observe their behavior, and learn from their experiences. Their engagement facilitates a unique opportunity to learn from a diverse range of role models and mentors, accelerating their own skill development. However, challenges arise when considering the potential drawbacks of this unfamiliar landscape.

AI platforms can sometimes reinforce negative behaviors and misconceptions, especially when algorithms prioritize engagement over educational value. Bandura's social learning theory has revolutionized educators' understanding of the learning process. Their three core concepts provide a framework for this futuristic classroom.

AI-powered systems that can record and replay lessons facilitate the first concept, allowing students to observe their peers and mentors, as well as others' behaviors. By considering each student's unique cognitive needs, personalized learning paths address the second concept, the role of internal mental states. The third concept emphasizes that learning doesn't always lead to change. The future classroom is about more than just knowing things; it's also about using them.

The students do not limit their attention to the information on their screens or holographic displays as they engage with their lessons. They are aware of their peers and the behavior exhibited around them. They witness their classmates' successes and failures, and, through Bandura's lens, they come to understand that learning is not just an individual pursuit but also a social one. This classroom is a microcosm of the larger world, where collaboration and observation are just as crucial as solitary study.

The Impact of AI and Technology on College Professors

Adult learners may need guidance to discern between accurate and misleading information and to develop critical thinking skills to evaluate the behaviors and information they encounter. This new classroom demands a certain level of digital literacy and discernment, skills that educators are now incorporating into their teaching practices.

The impact of AI and technology on college professors is twofold. On the one hand, they must adapt to the changing landscape by integrating AI into their teaching methods and staying up to date with the latest technological advancements. They can facilitate deeper learning and critical thinking by guiding students through the interpretation and application of information. This facilitation has two significant effects. Adults can learn in ways that suit them. There is also a risk of becoming overly reliant on technology and losing touch with traditional social learning methods.

Discussion forums play a pivotal role in this new educational landscape. They serve as virtual meeting places where students can discuss ideas, share insights, and learn from each other. In these online spaces, students witness how their peers think, solve problems, and express emotions, which enhances the observation learning theory.

This observation fosters a deeper understanding of the material and strengthens their social learning skills. Over time, as students observe their peers' behaviors reinforced through successful problem-solving, improved grades, and teacher recognition, they internalize these behaviors as their own.

Bandura's theory goes a step further, suggesting that individuals can also change their behavior by observing positive role models. Classroom students often seek out peers who excel in specific topics or possess effective study methods. They change their habits and attitudes to resemble those of their role models.

AI's Influence on the American Classroom

The influence of AI and technology on American classrooms is also significant. It has disrupted traditional methods, fostering a collaborative and immersive learning environment. Amidst these transformations, students find themselves at the center of a debate. The very tools that enhance their learning experiences could, some fear, also create an over-reliance, affecting their interpersonal skills and critical thinking abilities. Yet, these young scholars are eager to navigate this unfamiliar terrain, embracing the opportunities presented by AI while navigating the challenges with resilience and adaptability.

Students are no longer passive recipients of knowledge; they are active participants, teaching and inspiring each other. The future of education is here, a testament to the power of observational learning and the potential of our students. However, the disruption goes beyond just adopting new tools. The very nature of learning has developed.

Social learning theory is simple, as students not only observe but also discuss and reflect on their experiences with technology. They analyze their peers' behavior and outcomes, incorporating these insights into their own learning journeys. Emotional intelligence plays a pivotal role in this process, as students navigate the complex emotions that arise when interacting with AI and augmented reality. They learn to recognize and manage their emotions, developing resilience and empathy as they adapt to these new technologies.

The digital age has transformed not only the workplace but also the very foundation of education. The rise of the internet and AI has had a profound impact on the student experience and the traditional classroom environment. While it has brought about many opportunities, it has also presented its fair share of challenges.

The classroom has developed, and with it, the student experience. The internet and AI have revolutionized how we disseminate and absorb knowledge, and social learning theories are playing out in new and interesting ways. Students today have access to a wealth of information at their fingertips. Adults are witnessing a shift in their learning journeys, with AI offering unique opportunities and challenges.

The internet has become an indispensable tool, offering endless opportunities for learning and growth. With just a few clicks, students can explore diverse subjects, connect with experts worldwide, and engage in interactive learning experiences. However, this abundance of information has also led to challenges in critical thinking and information literacy. Students must now navigate a sea of misinformation and develop the skills to distinguish between reliable sources and fake news.

AI has also made its way into the classroom, offering innovative teaching and learning methods. AI-powered tools can personalize learning experiences, adapting to each student's unique needs and providing instant feedback. Yet, there is a concern that overreliance on technology may hinder the development of soft skills, such as interpersonal communication and emotional intelligence.

As millennials continue to take on leadership roles, the focus on soft skills becomes even more crucial. Finding a balance between technology integration and human-centric education is essential for preparing students for the future workplace. AI-powered tools present students with unprecedented opportunities. Interactive tutorials, adaptive learning algorithms, and virtual simulations engage students in ways traditional classrooms cannot.

AI can provide instant feedback, allowing students to learn primary concepts at their own pace. This level of personalization enhances learning outcomes and keeps students motivated. This reliance on AI in education also presents challenges. There are concerns about the potential negative impact on inter-personal skills and critical thinking abilities if students become dependent on AI tools. The rapid evolution of technology may leave some educators struggling to keep up, creating a skills gap in the classroom. Ensuring fair access to AI resources is another significant challenge in under-resourced communities, where students often lack the necessary infrastructure and digital literacy to engage with these new tools.

AI Tutors and Virtual Simulations

The classroom has indeed become a hub, but it extends beyond physical walls. Discussion forums play a pivotal role, serving as virtual extensions of the school. Here, social learning thrives as students collaborate, discuss, and debate ideas. These forums provide a platform for students to observe and analyze diverse perspectives and behaviors. They witness their peers' critical thinking, problem-solving, and emotional intelligence in action. The forums become a safe space for learners to practice and develop their own emotional intelligence, recognizing and managing their emotions as they navigate this digital landscape.

The value of soft skills extends beyond the classroom and into career development. As highlighted by Ghaffar et al., nurturing soft skills enables individuals to build social capital, which is essential for both personal and professional growth. Soft skills enhance one's ability to interact effectively, build strong relationships, and navigate social situations with confidence and ease. They are the cornerstone of successful collaboration and a key driver of career advancement. Bandura (1986) referred to this phenomenon as observational learning, emphasizing that individuals can acquire new knowledge and skills by observing and emulating others.

The future classroom, where students encounter various behaviors and reinforcement patterns, provides a perfect testing ground for this theory. They witness their peers engaging with AI tutors, collaborating on virtual projects, and interacting with augmented reality simulations. The students are agents of their own learning, navigating a classroom transformed by AI and technology. The students are at the heart of this revolution, which has disrupted the very fabric of education.

AI tutors have become integral, with an impact that is twofold. Not only do they engage and teach students, but their presence also facilitates observation learning. Bandura's theory is uncomplicated. As students witness their peers interacting with AI tutors, they emulate successful strategies and behaviors. These interactions create a ripple effect, with even the most reluctant students becoming engaged and their confidence growing as they observe and learn from their classmates.

The classroom has become a hub of collaboration, with students working together on virtual projects and teaching and inspiring one another. AI tutors have become a common sight, with students observing and learning from the interactions between their peers and these intelligent machines. In copying successful strategies, learners soon engage with AI tutors, and even the most hesitant students gain confidence and skills. A world of possibilities welcomes them as they step into this developing classroom. The once-static lectures have now been supplemented with interactive tutorials and virtual simulations.

AI tutors adapt to each learner's unique learning curve, offering a personalized journey through complex subjects. This dynamic environment keeps students engaged and motivated, rather than passively receiving information. However, with these advancements come challenges and essential questions.

As AI transforms the educational landscape, concerns arise about the potential digital divide and equity of access. Not all learners may have the resources or digital literacy skills to engage with these new tools. The rapid pace of technological evolution also raises concerns about communication and ethical dilemmas, leaving some educators struggling to keep up with the impact of AI and other technologies on American classrooms.

II

Part 2: AI and the Developing Learner

Part 2 effectively incorporates established adult learning theories, such as Knowles' andragogy and Kolb's experiential learning cycle, providing a solid foundation for analyzing the impact of AI on adult learning and communication. This section of the book maintains a consistent focus on AI's influence on communication, ethical considerations, and the adaptation of adult learners in the digital age.

It also offers practical solutions and strategies for addressing the skills gap, including industry partnerships, curriculum reform, agile learning methodologies, and AI-powered adaptive learning platforms. Overall, this section provides a comprehensive overview of AI's impact on communication and ethical considerations in adult learning.

3

AI's Influence on Communication and Ethical Considerations

The Fundamental Shifts in Communication

E volving communication has led to a shift in the way we interact and build relationships. While technology has gifted us with efficient, instant connections, it has also led to a detachment from traditional, personal methods of communication. The art of conversation and face-to-face interaction is becoming a lost skill, as we opt for the easier route of sending a quick message or commenting on a post. We now express ourselves through curated online profiles and craft our personal brands through digital footprints. This new form of self-expression affects our identity and how we wish to be perceived.

The online world has also changed the way we form and maintain relationships. We can now connect with people from across the globe with similar interests and form communities and friendships without ever meeting in person. The connection has created a sense of belonging and connection, but also a feeling of isolation.

We now spend more time on our phones and less time building real-world relationships. The impact of this shift on communication is far-reaching, with ripples in how we operate in both our personal and professional lives. As we navigate this unfamiliar landscape, it is essential to remember the value of human connection and the power of effective communication, both online and offline.

Today, communication encompasses everything from emails and social media to speaking, giving feedback, and creating rapport. Communication is a soft skill. Other soft skills include strategic thinking and interpersonal skills, which employers across all industries highly value.

Today, many social connections are online. Instead of joining a club, people often turn to apps or online groups for social connections. Instead of sitting down with someone and sharing, they share photos and news snippets on phones and tablets. This pervasive digital mediation is redefining what constitutes meaningful communication.

The emphasis has moved from extended dialogues and in-depth discussions to more frequent, yet often shallower, exchanges. This evolution affects not only personal relationships but also professional communication, where the ability to craft concise, impactful messages and maintain a strong online presence has become a critical asset. The digital world is now key to our social and professional lives. Access means that building connections and communicating successfully across industries requires adaptability and understanding of online etiquette.

Evolving communication in the digital age has been profound, with social media at the forefront of this transformation. Once a realm of in-person interactions and written correspondence, human connection has now expanded into the virtual world. Screens have become the alternative meeting places, and "lol" and "smh" are just a few of the many acronyms that have woven themselves into our linguistic fabric.

The very nature of how we express ourselves and forge relationships has undergone a fundamental shift. This fundamental shift is evident in the decline of traditional social structures, such as joining clubs for shared interests. Instead, platforms and applications now serve as the primary hubs for community formation and social interaction.

Communication in the Workplace

The act of broadcasting personal experiences through digital media has replaced sitting down with someone for a shared experience. This transition means that screens now mediate conversations, feedback, and the development of interpersonal skills, which alters the qualitative nature of these interactions and their implicit expectations. Employers, recognizing this evolution, seek candidates who can navigate these digital communication channels, alongside other crucial soft skills, such as strategic thinking and leadership.

The impact of this new communication landscape is clear in the workplace. As employers recognize the importance of prizing soft skills, they seek employees who can navigate both the digital and physical realms. Leaders and managers should possess a unique set of skills that blend traditional soft skills with digital proficiency. They must be able to build and maintain relationships with colleagues and clients, both online and offline, and adapt their communication style to different platforms and audiences. With a tap of a finger, we can now connect with people across the globe, sharing our thoughts, photos, and experiences.

Malcolm Knowles' andragogy, which emphasizes self-direction, experience-based learning, problem-centeredness, and intrinsic motivation in adult learners, offers a framework for understanding how individuals adapt to these new communication paradigms. A self-directed nature means that adults often explore and adapt to online environments through trial and error, driven by the need to remain connected and professional. Their experiences in face-to-face interactions now guide how they navigate virtual spaces, even though they must adjust their approach to building rapport and sharing information.

The problem-centered orientation encourages adults to understand the functional aspects of these digital tools, focusing on how they can achieve their communication goals. The fundamental shifts in communication, driven by the digital landscape, present a fertile ground for examining established adult learning theories.

37

Experiential Learning

David Kolb's theory of experiential learning, as presented in "Experiential Learning: Experience as the Source of Learning and Development" (1984), further illuminates the adaptive process. This theory outlines a cyclical progression involving concrete experience, reflective observation, abstract conceptualization, and active experimentation.

Applied to digital communication, individuals first encounter new platforms and interaction methods (concrete experience). They reflect on the outcomes of their digital exchanges, assessing the success or failure of the messages they convey (reflective observation). This reflection informs the development of new conceptualizations and strategies for online communication (abstract conceptualization), which are then tested and refined through subsequent digital interactions (active experimentation). This cycle is essential.

Adults need to learn the unspoken rules of digital communication and apply them effectively. Understanding these rules will improve their communication. David Kolb's experiential learning cycle provides a lens for analyzing the qualitative impact of AI on adult skill development.

AI-driven simulations and interactive platforms can provide rich, concrete experiences, allowing adults to engage with new concepts and scenarios. The subsequent reflective observation and abstract conceptualization stages become crucial as learners process the data and outcomes presented by AI, formulating their understanding and strategies.

David Kolb's experiential learning cycle provides a framework for analyzing the qualitative impact of AI on adult skill development. The cycle includes concrete experience, reflective observation, abstract conceptualization, and active experimentation.

AI-driven simulations and interactive platforms can provide rich, concrete experiences, allowing adults to engage with new concepts and scenarios. The subsequent reflective observation and abstract conceptualization stages become crucial as learners process the data and outcomes presented by AI, formulating their understanding and strategies.

Emotional intelligence, a key soft skill, has taken on an extra dimension in this digital age. IT professionals, for example, must navigate the complexities of human emotions through a screen. Understanding and interpreting digital body language, such as response times and emoji choices, has become an essential skill for fostering meaningful connections and collaborations. As our world continues to blend the digital and the physical, soft skills will remain at the heart of effective communication and leadership.

The effectiveness of AI-assisted learning then hinges on the adult learner's ability to translate these conceptualizations into active experimentation, integrating their newfound knowledge and skills into real-world applications. This cyclical engagement, amplified by AI's adaptive feedback, is crucial for consolidating learning and fostering adaptability in a rapidly evolving digital landscape.

The Era of Digital Technology

A revolution in self-presentation has come with the advent of the digital age. We craft our online personas, and each post, comment, and like contributes to our digital identity. This new form of self-expression enables us to showcase our interests, beliefs, and values to a global audience, but it also presents unique challenges.

As we spend more time curating our online brands, we risk neglecting the authentic connections and relationships that formed through face-to-face interactions. The ease and accessibility of online communication have led to a preference for virtual connections over real-world encounters.

The shift towards digital relationships can have consequences. The lack of physical presence and non-verbal cues in online interactions can lead to misunderstandings and miscommunications. Humans interact differently because of the digital world. Virtual presence has redefined how people connect and share.

Today, curated digital profiles, asynchronous messaging, and sharing visual content have supplanted these elements. This transformation has given rise to a new paradigm of communication. It prioritizes brevity, visual appeal, and quick sharing of personal news and updates, often using algorithms to boost engagement.

Andragogy emphasizes adults' interest in how digital tools facilitate their communication goals. Adults, driven by the need to remain connected, navigate unfamiliar digital platforms and communication styles through trial and error, a testament to their self-directed nature. Their prior experiences, shaped by face-to-face interactions, now inform their approach to virtual spaces, prompting a recalibration of rapport-building and information-dissemination strategies.

The impact of AI on adult learning and communication has been profound, offering new avenues for knowledge acquisition and interaction. As AI continues to develop, it has raised ethical concerns and questions about academic integrity in AI-assisted learning. AI has transformed the very nature of discourse and information, with additional considerations and challenges emerging.

Digital town squares have evolved into social media platforms, where people exchange ideas, start movements, and foster communities. They have empowered individuals to find their voice and enabled them to reach audiences once unimaginable. This unfamiliar landscape has also presented challenges.

The art of face-to-face conversation, with its nuances of body language and tone, is at risk of being overshadowed by the convenience of digital communication. Evolving communication in the digital age, driven by rapid technological advances and the ubiquitous presence of artificial intelligence, has transformed how we interact and connect. AI has reshaped the very nature of human communication, serving as both a catalyst and a conduit for alternative forms of expression and understanding. This shift has had a profound impact on the workplace in industries that rely on IT professionals.

Digital Technology in a Global Workplace

The digital age has also given rise to a global workplace, where diverse workgroups interact across borders and cultures. Practical communication skills are imperative for employees to build bridges and foster understanding within these diverse teams. AI has played a pivotal role in breaking down these barriers, providing translation services and cultural insights that were once challenging to access. As technology continues to advance, so too must our approach to communication, ensuring we harness the power of AI to strengthen our connections and empower future generations.

As employers seek candidates who can navigate this digital landscape, the onus falls on individuals to update their skill sets. Their efforts suggest a need for self-assessment and a willingness to engage in informal learning, including observing effective communication patterns online and experimenting with different approaches. The process of giving and receiving feedback, an essential element of effective communication, has evolved from direct verbal interaction to including written comments, emojis, and digital reactions on various online platforms. We must understand new feedback in light of the changing environment. Adults require refined communication to stay relevant in their lives.

New Platforms and Communication Styles

Adapting communication strategies by adults is a complex phenomenon intricately linked to established adult learning theories, which provide a framework for understanding this ongoing evolution. Malcolm Knowles' andragogy, a theory that posits distinct principles for adult learning compared to pedagogy (childhood learning), offers critical insights. Knowles's "The Adult Learner" (1984) notes that adults are self-directed learners who utilize their experiences. Knowles's work on andragogy offers a key perspective. Knowles characterizes adult learners by their self-direction, reliance on experience, problem-centeredness, and intrinsic motivation.

In digital communication, adults engage with new platforms and interaction styles, often through self-directed exploration and trial-and-error, driven by the imperative to maintain social and professional connections. Their experiences in face-to-face interactions now help them navigate virtual environments, requiring them to rebuild rapport and share information in different ways. The problem-centered orientation of andragogy is clear as individuals seek to grasp the functional aspects of digital tools, prioritizing their efficacy in achieving communication objectives.

The continuous adaptation required of adults in their communication strategies aligns with the concept of lifelong learning, particularly in soft skills. As employers prioritize candidates adept at navigating the digital realm, the responsibility for continuous skill development falls upon individuals. This prioritization leads to self-assessment and a commitment to informal learning, which involves observing effective online communication patterns and experimenting with various approaches to enhance one's skills.

Adults must understand how people process and interpret these alternative forms of feedback, as feedback mechanisms have evolved from direct verbal exchanges to digital comments, emojis, and reactions. Adults must refine their communication strategies to stay relevant in the digital world. Their understanding is crucial to both personal and professional success, underscoring the enduring value of adult learning theories. This ongoing refinement of communication strategies is paramount for adults to maintain relevance and effectiveness in both their personal and professional spheres within the dynamic digital ecosystem.

Adults adapt to their communication strategies. This adaptation aligns with lifelong learning, particularly in terms of developing soft skills. AI is becoming more critical in this area. As employers value candidates proficient in navigating the digital landscape, the onus of continuous skill development rests with individuals.

AI tools can facilitate this by offering personalized feedback on communication patterns, suggesting areas for improvement in digital etiquette, and providing access to vast libraries of learning resources. This interaction fosters a commitment to self-assessment and informal learning, encompassing the observation of effective online communication patterns and the adoption of various approaches, often guided by AI-driven insights.

The Ethical Considerations of AI-assisted Learning

Evolving AI has had a profound impact on the way adult learners acquire knowledge and engage with their peers and instructors. As AI continues to advance, it offers new learning opportunities but also presents ethical dilemmas and challenges to academic integrity. Communication and knowledge exchange have undergone significant transformations, altering the skills required of learners and educators.

With AI-assisted learning, adults can now access a wealth of information and connect with others more easily. However, this unfamiliar landscape also raises concerns about plagiarism and the potential misuse of AI. As technology advances, so too must educators' strategies to ensure academic integrity. These strategies include promoting ethical behavior and critical thinking, as well as providing guidance on the responsible use of AI tools and resources.

Emotional intelligence and collaboration skills are essential for adult learners, especially in IT professions. As technology plays an increasingly larger role in communication, the ability to navigate social situations and demonstrate empathy becomes even more crucial. IT professionals must possess these skills to ensure the success of digitization projects, meet user needs, and collaborate effectively with customers, stakeholders, and colleagues. AI has become an integral part of daily life for many adult learners, offering convenient and personalized ways to gain knowledge.

Virtual assistants provide tailored language lessons, and AI-driven search engines offer instant answers to queries, revolutionizing self-directed learning. AI also affects more traditional academic settings, with AI-assisted tutorials and adaptive learning platforms improving classroom teaching. However, alongside these advancements come ethical dilemmas. Students may misuse AI tools to plagiarize or generate false information, which is a primary concern.

As AI becomes more sophisticated, distinguishing between human-created and AI-generated content becomes challenging. These challenges blur the line of academic integrity and raise questions about the authenticity of knowledge. As AI continues to integrate into our daily lives, the demand for soft skills in the workplace has never been higher. Emotional intelligence, once considered a nice-to-have trait, is now a crucial component of successful collaboration.

Integrating AI into Learning & Development Environments

IT professionals, often working within independent silos, must navigate complex emotional landscapes and foster meaningful connections with colleagues and customers. This new reality demands a delicate balance between technical proficiency and a nuanced understanding of human interaction, a skill set that AI cannot replicate but can enhance.

Integrating AI into adult learning environments deepens exploration of how these technologies align with and challenge established adult learning theories. Malcolm Knowles' andragogy, which emphasizes the self-directed, experience-based, problem-centered, and motivated nature of adult learners, provides a foundational framework for understanding this interaction. AI tools offer personalized learning and accessible information. They support adult learners' need for self-direction, allowing them to learn at their own pace and meet their specific needs.

AI can also enhance the problem-centered orientation of andragogy by presenting real-world scenarios, which help adults gain practical skills applicable to their professional or personal challenges. Educators must address the ethical implications of AI. They are the custodians of this learning environment. Teachers and other educational professionals adapt not only to new pedagogical tools but also to a redefined responsibility of guiding learners toward critical engagement with AI-generated content, fostering a discerning mind rather than mere passive consumption.

The ethical considerations of AI-assisted learning are complex and far-reaching, demanding our attention as we navigate this new educational landscape. The concerns surrounding AI's impact on academic integrity are well-founded, especially as technology becomes increasingly sophisticated and accessible. Adult learners, with their unique needs and experiences, find themselves at the heart of this debate. On the one hand, AI offers unparalleled opportunities for personalized learning, filling the gaps that traditional education often leaves.

Virtual tutors, for instance, can provide tailored instruction and immediate feedback, catering to each learner's specific needs. This level of customization can accelerate learning and empower adults who may have struggled in traditional educational settings. However, the potential for misuse is ever-present. As AI tools become more advanced, the temptation to rely on them, or even to abuse their capabilities, grows. AI might tempt some learners to generate entire essays or research papers, thereby compromising the essence of academic integrity.

The ease of accessing and manipulating information challenges the traditional notions of learning and raises questions about the authenticity of the knowledge gained. In this developing landscape, establishing ethical guidelines and educational frameworks that emphasize responsible AI usage is crucial. Institutions and educators must collaborate to raise awareness about the proper use of AI tools and the potential consequences of their misuse.

Fostering Academic Integrity

By fostering a culture of academic integrity that embraces the benefits of AI while recognizing its limitations, adult learners can navigate this unfamiliar terrain with confidence and integrity. The ongoing development and influence of AI on how we gain and communicate knowledge threatens our adherence to the fundamental principles of academic integrity.

Developments include fostering critical thinking and encouraging learners to question and analyze the information they encounter. By doing so, we empower them to distinguish genuine understanding from mere regurgitation of information, whether generated by humans or machines. Emotional intelligence plays a crucial role in this digital era.

As adult learners navigate a virtual learning environment, they must be able to interpret and respond to digital body language, such as response times and emoji use. This new form of nonverbal communication affects the way we collaborate and connect with our peers. It requires a unique set of skills to foster meaningful interactions and build a sense of community, even when we are not present. In this evolving educational landscape, it is crucial to address these ethical concerns and adapt our strategies to ensure the responsible use of AI.

Integrating AI into adult learning environments deepens exploration of how these technologies align with and challenge established adult learning theories. The impact of this shift on communication is far-reaching, with ripples that have affected how we operate in both our personal and professional lives.

Educators must play a pivotal role in guiding adult learners to question the accuracy and bias of information, foster critical thinking, and promote ethical behavior. By doing so, we empower learners to navigate this complex digital world with confidence, integrity, and a nuanced understanding of AI's impact on their learning journey.

Fostering Adaptability in an Evolving Digital Landscape

AI facilitates the acquisition of practical skills applicable to an adult's professional or personal challenges, thereby improving the problem-centered orientation of andragogy. This technological augmentation, however, requires a mindful approach to ensure that the learning process remains grounded in genuine understanding rather than superficial information retrieval. The observed shift in communication dynamics presents an interesting intersection with established theories of adult learning.

The impact of AI on adult learning has also changed the way educators approach their roles. Teachers and instructors must now blend traditional pedagogical methods with a new understanding of digital tools and their potential pitfalls. Their efforts include guiding learners to question the accuracy and bias of information, while also encouraging them to consider the role of AI in knowledge creation and dissemination.

The growing role of AI and the developing nature of communication have significantly impacted our personal and professional lives. They've also changed how we understand knowledge and academic honesty. As we navigate this digital age, the very nature of human interaction and self-expression continues to undergo a metamorphosis.

Screens and devices have extended our identities, with our online personas curated to present ourselves to the world. This shift has had a notable impact on adult learners, who now have access to a vast array of AI-assisted learning tools and resources. The convenience and personalization offered by AI have revolutionized self-directed learning.

Adult learners can now access virtual assistants for language lessons, instant answers to queries through AI-driven search engines, and tailored instruction from virtual tutors. However, with these advancements come ethical dilemmas regarding academic integrity and the potential misuse of AI. As AI becomes more sophisticated, distinguishing between human-created and AI-generated content becomes challenging, blurring the lines of authenticity and raising concerns about plagiarism and the integrity of knowledge acquisition.

The effectiveness of AI-assisted learning then hinges on the adult learner's ability to translate these conceptualizations into active experimentation, integrating their newfound knowledge and skills into real-world applications. This cyclical engagement, amplified by AI's adaptive feedback, is crucial for consolidating learning and fostering adaptability in a rapidly evolving digital landscape. The educator's role changes in this context. The rapid pace of technological evolution also raises concerns about a skills gap, leaving some educators struggling to keep up with the impact of AI and other technologies on American classrooms.

4

Emerging Industries and Soft Skills Gaps

Skills Gap Crisis

T he growing disparity between the skills graduates possess and those employers seek is a pressing issue. Both public and private educational institutions are struggling to bridge this divide, leaving students ill-prepared for the demands of the workplace. The skills gap is especially noticeable in science and technology. Rapid advancements in these fields require a dynamic and adaptable skill set.

The deficiency of industry-relevant soft skills among graduates is a critical factor contributing to this challenge. Employers seek competencies such as effective communication, problem-solving, and critical thinking. However, academic curricula do not always integrate them.

A misunderstanding between educational outputs and industry require-ments in the science and technology sectors presents a challenge. This skills gap does not remain static. Graduates often lack the adaptability, problem-solving, and collaboration skills that employers value. Further compounding this issue is the rapid evolution of industry demands themselves. What constitutes a relevant skill today may become obsolete within a few years, necessitating continuous adaptation from both educational institutions and individuals.

Predictive modeling of future skill requirements is an ongoing endeavor, but the inherent unpredictability of technological innovation makes it a complex undertaking. The skills gap crisis has left both educators and employers searching for solutions. While the problem is multifaceted, some innovative approaches are showing promise in bridging the divide.

One such approach is integrating industry partnerships into the academic curriculum. The skills gap between graduates and industry demands persists, leaving graduates at risk of being unemployable. People often misunderstand soft skills, viewing them as unessential in the digital age.

Skills Mismatch

Adult professionals, aware of the evolving nature of work, feel unease as they recognize that the soft skills they possess may not be sufficient for the future. The gap is not a perception but a pressing reality. As industries undergo digital transformation, soft skills such as critical thinking, problem-solving, and effective communication are becoming valued. However, many graduates lack these very skills, raising concerns among employers. This mismatch between the skills employers seek and those graduates possess is a critical issue that demands attention. Educational institutions must address this deficit in soft skills.

By integrating industry-relevant soft skills training into their curricula, they can better prepare graduates for the demands of the modern workplace. Bridging this gap will ensure that students possess not only the technical knowledge but also the soft skills necessary to thrive in their chosen fields. The industry demands skills that schools often do not teach, creating a pressing issue, especially in information technology.

The rapid evolution of hardware and software development has created a dynamic and challenging landscape for IT professionals. Schools aim to equip graduates with essential hard skills, such as coding and technical knowledge, and it's also crucial that they prioritize soft skills. Graduates may possess strong technical skills due to a misunderstanding between academia and industry, but they may lack the ability to apply their knowledge in the changing business environment.

Encouraging a culture of lifelong learning is essential to bridge this gap. Someone must provide IT professionals with resources and incentives to update their knowledge and skills. These efforts could involve partnerships between businesses and educational institutions, offering employees access to the latest research and training in emerging technologies.

Companies would benefit from fostering an environment that values and rewards the development of soft skills. By recognizing and valuing effective communication, creative problem-solving, and critical thinking, organizations can empower their employees to fill skill gaps and adapt to the industry's dynamic nature.

Graduates entering the IT field often struggle to navigate the complex, ever-changing demands of the industry. While their technical prowess may be impressive, with a firm command of coding and hardware knowledge, it is in applying these skills that the gap becomes apparent. The dynamic nature of the IT sector requires professionals who can adapt, innovate, and use their knowledge effectively—and this is where soft skills become essential. Adult professionals in information technology are feeling the impact of this skills gap. They find themselves in a predicament where their technical abilities, although necessary, are insufficient to keep pace with industry's evolving demands.

The gap between the soft skills they possess and those required to maintain employment in the digital age is widening. Misunderstanding does not just affect the graduates. The belief that soft skills are a nice-to-have rather than a necessity is changing. Industry experts and employers are now seeking professionals who not only possess technical know-how but also soft skills to thrive in a dynamic business environment. These skills include effective communication, problem-solving, critical thinking, and the ability to work.

Continuous Learning

The ongoing challenge of bridging the skills gap deepens understanding of adult learning principles. The challenge of bridging the skills gap amplifies when considering the adult learning theories that underpin effective workforce development. Malcolm Knowles's theory of andragogy provides a robust framework for conceptualizing effective workforce development. Andragogy posits that adult learners are self-directed and draw upon a rich reservoir of life experiences. A clear understanding of the need to learn motivates them, and they are most receptive when faced with a problem or a practical need for new knowledge.

For recent graduates entering the professional arena, recognizing their skill deficiencies serves as a potent catalyst for learning, aligning with the core tenets of andragogy, which emphasize self-concept and readiness to acquire new competencies. Adult learning emphasizes practical, problem-centered approaches. Therefore, training interventions should focus on real-world industry challenges to address soft skill gaps. Bridging this widening chasm requires a multifaceted approach that extends beyond curriculum reform. Industry partnerships are proving crucial in identifying and imparting these emerging skill sets.

The aim is to equip graduates not only with the knowledge but also with the practical and interpersonal skills that foster employability and long-term career success. Their ability will measure the effectiveness of these strategies in preparing graduates to meet the current and future demands of a dynamic global marketplace. Educational institutions and industry partners play a vital role in facilitating this transformation by cultivating environments that encourage critical discourse, peer feedback, and the exploration of diverse perspectives.

Graduates should develop a mindset of continuous learning and adaptability. This adaptation will help them meet current industry demands. It will also enable them to anticipate and adapt to future changes in science and technology, thereby addressing the skills gap. The rapid pace of technological change demands continuous learning and development; yet universities often struggle to offer accessible, relevant programs for working adults. This shortcoming hinders professionals from enhancing their technical skills and adapting to emerging industry trends, thereby affecting their career prospects and contributing to a shortage of skilled labor.

Students learn about new technologies and methods through training, research, and teamwork-based work programs. The adoption of agile learning methodologies within educational institutions, which mirrors the iterative nature of technological development, enables the rapid integration of new content and skill-building opportunities. It also ensures graduates have foundational knowledge, allowing them to adapt with foresight to meet the demands of the modern and future job market.

The IT industry is evolving. Closing the skills gap is crucial to building a future-ready workforce that can meet changing digital demands. IT professionals are in high demand across sectors, and their roles have expanded beyond simple data encoding and report generation. Businesses now rely on IT for process outsourcing, and this requires a diverse set of skills.

Soft skills, such as communication, problem-solving, and adaptability, are crucial for IT professionals to collaborate effectively with colleagues, comprehend complex business needs, and deliver innovative solutions. However, the current education system often falls short of equipping graduates with these skills, creating a mismatch between what employers seek and what graduates offer. The gap harms both graduates' employability and the broader productivity and competitiveness of industries.

New and Evolving Industries

As demand for IT professionals continues to grow, educational institutions must recognize the field's evolving nature and adapt their curricula accordingly. By balancing the development of hard and soft skills, schools can produce well-rounded graduates who are better equipped to meet the diverse challenges of the IT industry. Emerging new sectors, driven by rapid technological innovation, further complicate the existing skills gap. Consider, for instance, the burgeoning fields of artificial intelligence ethics, quantum computing development, and sustainable biotechnology. These sectors, conceptualized a decade ago, now demand specialized skill sets that traditional educational pathways have yet to integrate.

Graduates entering these nascent industries often need to gain new competencies, such as proficiency in AI bias detection, quantum algorithm design, or bio-nanotechnology fabrication. Because of this, we must continually update our teaching and expectations of future skills. The renewable energy sector requires individuals skilled in grid modernization, advanced energy storage solutions, and the integration of distributed energy resources.

The ability to pivot and acquire new technical proficiencies, coupled with strong critical thinking and adaptive problem-solving skills, becomes paramount. It creates a dynamic environment where lifelong learning is not an advantage but a fundamental requirement for sustained employability and career progression.

Labor Shortages & Recruitment Challenges

The labor shortage and recruitment challenges faced by employers in the U.S. within the science and technology sectors correlate with the identified skills gap. Companies are reporting significant difficulties finding candidates with the precise combination of technical proficiency and essential soft skills required to navigate the complexities of modern industry. This deficit is not a matter of insufficient technical knowledge; it extends to the critical competencies of problem-solving, adaptable thinking, and effective collaboration.

The inability to locate candidates with this holistic skill set leads to prolonged recruitment cycles, increased hiring costs, and a direct impact on organizational productivity and innovation. Businesses often need to invest in upskilling or external training to address skills gaps among new hires. It highlights the importance of matching education with industry needs. The rapid pace of technological advancement, which reshapes the demand for specific skills, amplifies this recruitment difficulty.

Technical skills valued by employers in advanced manufacturing, biotechnology, and artificial intelligence may be insufficient. These sectors seek those who can learn and adapt. The reliance on traditional academic curricula, which may not always keep pace with these rapid industry shifts, means that graduates often enter the workforce knowing they require immediate professional development.

Recruiters face a dual challenge: finding candidates with potential and then training them in the skills needed. This dynamic creates a persistent tension between the supply of qualified labor and the developing needs of the U.S. workforce. U.S. employers, particularly in the science and technology sectors, face persistent labor shortages and recruitment challenges linked to the skills gap.

Companies report significant difficulty identifying candidates with the precise combination of technical proficiency and essential soft skills needed to navigate the complexities of modern industry. This deficit extends beyond mere technical knowledge, encompassing critical competencies such as problem-solving, adaptable thinking, and effective collaboration.

Technical skills that were valuable just a few years ago are now often deemed inadequate by employers in fields such as advanced manufacturing, biotechnology, and artificial intelligence. They are seeking people who can learn new things and adjust to new methods. The aim is to cultivate a talent pipeline better prepared to meet current and future industry requirements.

This forward-thinking strategy acknowledges that both schools and businesses must collaborate to address the shortage of skilled workers. Employers are working to solve the labor shortage and hiring difficulties by investing in workforce development. They aim to maintain a steady supply of qualified workers who can drive innovation and maintain competitiveness.

Kolb's Experiential Learning Cycle, which involves concrete experience, reflective observation, abstract conceptualization, and active experimentation, provides a model for designing such learning interventions. In this context, experiential learning, a cornerstone of andragogy, becomes paramount. The technological gap misunderstanding exacerbates the gap, where universities lag in providing upskilling opportunities for adult professionals.

Crucial Element for Long-term Employability

Graduates can build knowledge and skills through these cycles. This approach is more effective than simply receiving information. It helps them adapt to the changing needs of science and technology. The focus shifts from gaining knowledge to developing the capacity to learn and apply it in novel situations, a crucial element for long-term employability. Therefore, it's vital to be forward-thinking, develop fundamental skills that enable individuals to learn and adapt, and avoid teaching only outdated, specific skills.

Jack Mezirow's Transformational Learning Theory offers another relevant lens for viewing effective adult learning. This theory highlights that adult learning is most impactful when it involves transforming one's existing ways of understanding. For graduates, this may entail inspecting their preconceptions about their own capabilities and the nature of professional work, alongside a thorough reflection on their accumulated experiences and inherent assumptions.

Experiential learning, a fundamental component of andragogy, emerges as a critical strategy in this context. Opportunities such as internships, apprenticeships, and project-based learning integrated within academic settings can furnish graduates with invaluable practical applications of theoretical knowledge. These experiences help foster the development of essential soft skills, including collaboration and problem-solving, which are areas of deficiency.

The Ever-Developing Nature of Information Technology

The ever-developing nature of information technology presents a unique challenge for professionals in the field. The rapid advancement of hardware and software development has widened the skills gap, requiring IT experts to adapt and acquire new competencies. Over time, the role of IT professionals has expanded far beyond simple data encoding and report generation. Information technology has become integral to businesses through various forms of outsourcing processes. This evolution has brought about new challenges and heightened demand for soft skills.

As organizations rely on AI for various tasks, a skills gap emerges in understanding and using new technologies. While AI can perform repetitive and time-consuming tasks, the human ability to analyze, interpret, and make complex decisions remains vital. The deficiency lies in the overdependence on AI interaction, which may hinder the development of critical soft skills. Industries that are quick to adopt new technologies may find themselves at a disadvantage if they do not also invest in training their employees in soft skills.

The ability to communicate, solve problems, and think is essential for any organization's long-term success. As such, businesses must balance harnessing the power of AI with nurturing their workforce's soft skills. It includes fostering creative thinking, emotional intelligence, and complex problem-solving, ensuring employees can adapt to the dynamic nature of the IT industry and address potential skill gaps.

Recognizing the deficit in graduates' employability skills, academic institutions are now offering courses and workshops focused on communication, problem-solving, and critical thinking. These programs aim to equip students with the tools they need to navigate the dynamic professional landscape and adapt to future challenges. Forward-thinking companies are now collaborating with educational institutions to provide students with hands-on experience and insight into the skills required in the modern workplace.

These partnerships offer students the opportunity to apply their knowledge in a practical setting and develop in-demand soft skills. However, the technological gap remains a significant hurdle. To address this, universities are reevaluating their technological infrastructure and pedagogical methods. Investing in state-of-the-art equipment and fostering a culture of digital literacy ensures that students have access to the latest tools and are proficient in their use.

Another strategy to gain traction is the implementation of dedicated soft skills training programs within universities. Internships, apprenticeships, and project-based learning can help graduates apply their knowledge. These opportunities also foster vital soft skills, such as collaboration and problem-solving. The development of online learning platforms and part-time, flexible programs caters to the needs of adult professionals, allowing them to upskill without sacrificing their careers.

The Potential of Artificial Intelligence (AI)

The potential of artificial intelligence (AI) to address the skills gap is a developing area of exploration. AI-powered adaptive learning platforms, for instance, can tailor educational content and pace to individual learner needs, a direct application of andragogy's self-directed and problem-centered principles. These systems can identify specific knowledge gaps and provide targeted resources, enabling graduates to acquire new competencies more efficiently.

AI can simulate complex industry scenarios, offering immersive experiential learning opportunities that complement traditional internships and project-based learning. It enables the practice of problem-solving and collaborative skills in a controlled environment, thereby enhancing the application of theoretical knowledge derived from abstract conceptualization within Kolb's learning cycle.

AI helps predict future skill needs. It analyzes industry trends and technological advancements to forecast emerging competencies, informing curriculum reform and the development of training programs. Industries like project management and education will also leverage AI. AI automates routine project management tasks, such as resource allocation and risk assessment. It allows human project managers to concentrate on strategy, communication, and complex problem-solving. Predictive AI models can analyze historical project data to forecast potential bottlenecks and optimize timelines.

Industries such as project management and education are well-suited for AI to address new skill needs and help adult professionals enhance their soft skills. AI's application extends to augmenting human capabilities, rather than replacing them. In fields that require advanced cognitive skills, AI can handle data analysis and pattern recognition, enabling human professionals to focus on higher-level reasoning, critical thinking, and ethical decision-making.

In education, AI can analyze student engagement patterns to provide instructors with valuable insights, enabling them to adapt their teaching strategies more effectively. AI and human intellect work together to create a more flexible and responsive approach to learning and skill development.

This partnership addresses the skills gap by enabling individuals to acquire and apply new knowledge in diverse situations. It promotes long-term employability and career success.

AI-driven tutoring systems can offer personalized learning paths. These systems identify areas where students struggle and provide focused support, aligning with the andragogy principle that people are ready to learn when they have identified a problem. AI can also assist educators in curriculum development by analyzing learning outcomes and identifying practical pedagogical approaches.

AI-powered adaptive learning platforms, for instance, can tailor educational content and pacing to individual learner needs, embodying andragogy's principles, which emphasize self-direction and problem-centered learning. These systems identify specific knowledge deficiencies, providing targeted resources that enable graduates to acquire new competencies more efficiently.

AI automates project management tasks, such as resource allocation and risk assessment. It frees human project managers to focus on strategy, communication, and complex problem-solving. It increases the need for strong communication and critical thinking skills. Predictive AI models can analyze historical project data to forecast potential bottlenecks and optimize timelines, reinforcing the importance of adaptive decision-making skills.

AI can help predict skills by analyzing extensive data on industry trends and emerging technologies, enabling us to update our teaching methods and create new training programs. Educational institutions and industry partners can collaborate by forecasting emerging competencies. It will allow them to make learning interventions. These interventions equip individuals to gain new skills and adapt to changing environments, thereby preventing the skills gap from widening.

In education, AI analyzes student engagement to provide instructors with valuable insights. It allows instructors to adapt teaching strategies, requiring strong analytical and communication skills. This division of labor enables a more effective utilization of both machine intelligence and human soft skills.

AI automates routine tasks and offers advanced analytical support. It allows individuals to function at a higher cognitive level. It enhances their ability to adapt to the changing demands of science and technology. It also helps them contribute more meaningfully to their organizations. Constant upskilling in soft skills remains crucial.

The Reliance on AI and the Subsequent Skills Gap

Applying AI also extends to augmenting human capabilities, rather than replacing them. Cybersecurity utilizes AI to analyze data, identify anomalies, and expand the capabilities of human experts. These experts can then focus on strategy, complex problem-solving, and ethical concerns. This division of labor enables a more effective utilization of both machine intelligence and human soft skills, such as critical thinking and effective communication.

The reliance on AI and the subsequent skills gap are pressing issues that demand attention from businesses and educational institutions alike. As industries develop, so must the methods of equipping professionals with the tools to navigate this landscape. A holistic approach to training is required, one that recognizes the importance of both technical and soft skills. While technical skills may provide the foundation, soft skills are the scaffolding that enables professionals to adapt, innovate, and lead.

III

Part 3: The Future of Learning & Development & Education

Part 3 highlights the significance of lifelong learning and soft skills in the contemporary workforce. It provides a clear and logical argument for the benefits of continuous learning for both individuals and organizations. This section of the book also includes references to external sources (SEMCOG, World Economic Forum, Schirf & Spiegel, Borner et al.) to support its claims, enhancing credibility. It concludes with a speculative vision for the classrooms of 2035, where decentralized learning hubs, AI tutors, and virtual reality simulations will create a new educational paradigm.

5

Lifelong Learning and the Most In-demand Soft Skills

A Skilled Workforce

By fostering an environment where continuous learning is the norm, employers can empower their workforce not only to meet current job requirements but also to anticipate future skill needs, enhancing both individual employability and overall organizational resilience. The impact of this symbiotic relationship between lifelong learning and a comprehensive framework of soft skills extends beyond individual career trajectories to societal economic performance and quality of life.

It is also impossible to overstate the importance of prioritizing lifelong learning, acquiring soft skills, considering career development, and navigating the competitive job market. Adopting a lifelong learning practice is suitable for both employers and employees, providing a structured approach to identifying and developing essential abilities. Lifelong learning encourages individuals to adapt to the changing needs of the job market. It also helps them develop skills such as communication and problem-solving. The World Economic Forum (2020) identifies these as crucial skills.

Having the soft skills to adapt and thrive allows individuals to contribute to a more dynamic and innovative economy. SEMCOG (2012) and the World Economic Forum (2020), among others, have researched the return on investment in a skilled workforce. This investment has a ripple effect, boosting the community. It increases productivity, citizen engagement, and the standard of living. Life-long learning is essential for society.

There is a plethora of soft skills categories that are fostered by work-force development professionals (SEMCOG, 2012). Workforce development professionals play a pivotal role in promoting a wide range of soft skills. By recognizing the unique needs of each individual and providing tailored guidance, these professionals help employees develop transferable skills across industries and positions. It empowers employees to adapt to changing market demands and position themselves for long-term career success. The impact of such an approach extends beyond the individual, as enhanced soft skills can lead to improved economic performance and a higher quality of life for entire communities.

Research underscores the far-reaching benefits of this approach. Employee continuous learning and development of soft skills has a significant impact on society. It enhances not only an individual's career trajectory but also their overall well-being and life satisfaction. It can lead to a more productive, innovative, and resilient workforce, capable of driving economic growth and societal progress. Thus, the soft skills framework, coupled with a commitment to lifelong learning, becomes a powerful catalyst for positive transformation in employees' lives and the communities they inhabit.

Lifelong learning provides an indispensable foundation for adult learners to identify and acquire the soft skills needed to maintain employment and advance their careers. The highest-quality research can have a substantial beneficial impact on society. It can improve economic performance and en-hance the quality of life. However, as the SEMCOG report from 2012 highlights, the current landscape is fragmented, with various soft skills categories often targeting specific audiences, limiting their reach and effectiveness.

The effectiveness of this framework increases when integrated with adult learning principles. The challenge, then, is to develop a comprehensive and inclusive framework that addresses the diverse needs of employers and employees across various industries. It entails raising awareness about the importance of soft skills and providing accessible pathways for adult learners to develop them. By doing so, individuals can enhance their employability, adapt to changing market demands, and improve their career prospects.

Nonconforming to Lifelong Learning

Individuals who opt out of lifelong learning in the realm of soft skills face a challenging employment landscape. Without continuous engagement in skill development, the foundational relevance of their existing proficiencies erodes. It can manifest in significant financial challenges, as employers prioritize candidates who show adaptability and a commitment to developing competencies.

The inability to articulate and show crucial soft skills, such as problem-solving, collaboration, and critical thinking, impedes an individual's ability to secure stable employment. Opportunities that require nuanced human interaction or strategic foresight become inaccessible, narrowing the job market and increasing the likelihood of periods of unemployment or underemployment.

The financial repercussions of this lack of continuous learning are multi-faceted. Beyond the immediate impact of job instability, individuals may find themselves excluded from roles offering competitive salaries and benefits. Companies that invest in their workforce through development programs are more likely to reward employees who reciprocate this commitment by upskilling.

Those who remain static in their learning journey risk being relegated to lower-paying positions or temporary contracts, creating a cycle of financial precarity. The difficulty in obtaining stable employment exacerbates an employer's perception of an individual as less adaptable and therefore a greater long-term risk.

During the hiring process, the ability to articulate recent learning experiences and future development goals often forms this perception, which affects an individual's capacity to meet market demands. A failure to commit to lifelong learning in soft skills can widen the gap between an individual's capabilities and the evolving needs of the job market. It creates not only sustained difficulty in obtaining but also in maintaining employment, leading to persistent financial strain and a diminished quality of life. Principles of andragogy indicate that adult learners find relevance and practical application to be motivating.

It's not that soft skills are scarce, nor is there a lack of development guides. The key is to distribute and apply these principles in a tailored way. To succeed, we must connect theory to practice, enabling individuals to use their skills daily and throughout their careers. Workforce development professionals, armed with the insights from high-quality research, must act as architects of this continuous learning journey.

Workforce development professionals must translate the broad strokes of a comprehensive soft skills framework into practical, accessible learning experiences that resonate with the diverse motivations and backgrounds of adult learners. Project managers and customer service reps need to develop different soft skills. Leadership is key for the former, while conflict resolution is key for the latter.

Remaining Relevant and Competitive

The concept of continuous learning is not merely a luxury in today's evolving job market; it is a necessity. Adults must embrace lifelong learning not only to keep pace with the changing demands of their industries but also to stay relevant and competitive. Acquiring soft skills, such as critical thinking, problem-solving, and effective communication, becomes the cornerstone of their journey towards career advancement and job security. Many people recognize the value of self-improvement. They seek tools and strategies from quality research to navigate the uncertain economy.

Research in this field has a significant impact on the role of soft skills in the modern workplace. By continuing to explore and innovate in this area, society stands to benefit from improved economic performance and an enhanced quality of life for its citizens. It is a journey of continuous improvement. Individuals with the right skills can navigate the dynamic employment landscape with resilience and adaptability. Rapid technological advancements drive an ever-changing world, making continuous learning imperative for adults to stay relevant in the job market.

The digital age has brought about a revolution, with old skills becoming obsolete at an unprecedented pace. To navigate this evolving landscape, individuals must adopt a lifelong learning approach and continually enhance their skills. It is especially true for soft skills, which employers seek.

The world is witnessing rapid technological advancement. In this dynamic landscape, the concept of continuous learning has emerged as a vital strategy for professionals seeking to remain relevant and competitive. Lifelong learning is no longer a choice but a necessity, as individuals strive to navigate the ever-changing job market and secure their employability. Acquiring new skills, particularly soft skills, is crucial for success in the modern era.

The societal impact of the collaboration between lifelong learning and the development of robust soft skills is substantial. Continuous adaptation and the acquisition of transferable skills empower individuals to contribute to a more dynamic and resilient economy. Research, such as the SEMCOG (2012) strategy and the World Economic Forum's (2020) outlook demonstrates that a skilled workforce yields economic benefits. This investment yields a ripple effect, improving community well-being by increasing productivity, enhancing civic engagement, and raising the overall standard of living.

Learning & Development

Lifelong learning, combined with soft skills, is crucial to societal progress and a higher quality of life. It is not just for individual career growth. Soft skills, such as communication, problem-solving, and critical thinking, are essential for success in the modern workplace. As technology automates and replaces routine tasks, the human ability to innovate, adapt, and interact becomes valuable.

Employers seek individuals with strong soft skills who can lead diverse teams, think, and solve complex problems. By investing in their personal and professional development, adults can enhance their employability and stay resilient in the face of technological disruption.

Those with strong soft skills can lead and inspire diverse teams, fostering a culture of collaboration and adaptability. By investing in personal and professional development, adults enhance their employability and build resilience, ensuring they remain valuable assets in a turbulent job market. The success of this integrated approach hinges on creating a culture where continuous learning is not just encouraged but embedded within the organizational fabric.

Employers who support the development of soft skills through accessible frameworks and a commitment to lifelong learning are investing in their most valuable asset: their people. This investment yields returns beyond immediate productivity gains. It cultivates an engaged, innovative, and resilient workforce prepared for future challenges. It will improve the quality of life for everyone.

When individuals learn and grow, the positive impact grows, creating progress throughout our communities. Workforce development professionals —the architects of this continuous learning journey —must translate the abstract principles of a comprehensive soft skills framework into tangible, actionable learning experiences. The aspiring project manager requires a distinct teaching approach compared to the customer service representative. Both want to improve their soft skills, but their goals differ.

Theory and practice must be linked; it's vital. It transforms skills into practical tools, empowering individuals and shaping their careers. The evolving nature of the workplace has led to a significant shift in the skills employees need to remain competitive.

Soft skills, including interpersonal skills and social acumen, are increasingly sought by employers. Schirf & Spiegel (2017) and Borner et al. (2018) have documented this trend over the past 20 years. They also emphasize the increasing significance of soft skills in professional settings. As individuals embark on their journeys of continuous learning, they soon realize the impact it has on their personal growth.

Soft skills are transformative. They boost confidence and resilience in both professional and personal life. They are better at making choices, solving problems, and approaching life with a desire to learn. By integrating the foundational tenets of andragogy, these dedicated professionals can cultivate learning environments that are not only relevant to the adult learner but also self-directed. It empowers adults to identify and address their unique skill deficits and pursue developmental pathways that align with their career aspirations. This transformation turns the soft skills framework into a dynamic roadmap. It guides individuals through the workplace, giving them the resilience they need.

Skillset Versatility

The future belongs to those with a versatile skill set, capable of adapting to the dynamic demands of a digital economy. Soft skills, such as emotional intelligence and critical thinking, will become increasingly valuable, complementing technical expertise. Professionals who can straddle multiple disciplines and bring a diverse range of skills to the table will be in high demand. Employees need to invest themselves in this to stay ahead, requiring self-investment, the acquisition of new skills, and adaptation to the shifting digital world. Organizations also play a pivotal role. They must create a culture that promotes the ongoing Nature of work.

The evolving nature of the workplace has led to a shift in the skills employees need to possess to remain competitive. Soft social skills have become a key differentiator in the employment landscape. According to industry experts, the ability to navigate social interactions and exhibit emotional intelligence is becoming just as important as technical proficiency.

Leaders must not only manage facilities but also facilitate board meetings and convey complex ideas clearly. It underscores the rising demand for soft skills in the workplace. Borner et al.'s research supports this notion, highlighting that over the last two decades, soft skills have gained prominence across various industries.

Employers are seeking candidates who can communicate, collaborate, and adapt to diverse work environments. These skills enable employees to build strong working relationships, resolve conflicts, and navigate the complexities of modern workplaces. As such, soft skills have become a crucial factor in hiring decisions and career advancement, with employers recognizing their impact on team dynamics, productivity, and overall organizational success.

A New Strategic Approach

Navigating this complex and dynamic environment requires a strategic approach. Individuals must seek out learning opportunities, whether through online platforms, workshops, or mentorship programs. By doing so, they equip themselves with the tools necessary to future-proof their careers. The focus should not be solely on technical skills but also on soft skills —the human abilities that distinguish one individual in a world of automation. Employers who seek individuals capable of leading diverse teams and innovating in the face of challenges value communication, problem-solving, and critical thinking.

People must continually learn and grow to keep pace with the world around them. It may involve enrolling in online courses, attending workshops, or taking part in industry conferences. By upgrading their knowledge and skills, professionals can future-proof their careers, adapt to changing market demands, and seize new opportunities. Lifelong learning is no longer an option but a key to survival and success in the modern era. The concept of continuous learning is not just a choice but a necessity in the modern era.

As technology advances, the job market will continue to undergo constant flux, with skills becoming obsolete faster than ever. Those who embrace lifelong learning will thrive in this developing landscape, adapting to industry shifts and seizing the opportunities presented by the digital age. This journey of self-improvement is empowering, fostering agility, competitiveness, and resilience in a transforming world. Individuals must seek learning opportunities, whether through online courses, workshops, mentorship programs, or personal initiatives.

By continually acquiring new skills and knowledge, professionals can future-proof their careers, adapt to industry shifts, and capitalize on emerging opportunities. This journey of continuous self-improvement empowers individuals to stay agile, competitive, and resilient in a changing economic landscape.

Investing in personal and professional development is a commitment to one's future. It enables individuals to remain resilient in the face of technological disruption and enhances their employability.

By embracing new skills and knowledge, professionals can adapt to evolving market needs and secure their place in it. This continuous journey of self-improvement is a testament to one's adaptability and determination, ensuring they remain relevant and competitive in a crowded and ever-changing job market.

AI and the Fourth Industrial Revolution

The rapidly evolving job market, driven by technological advancements, demands that adults embrace a lifelong learning approach. The digital age has brought about a revolution in which skills can become obsolete in the blink of an eye. To stay relevant, individuals must continually upskill and adapt, ensuring they remain valuable assets in a rapidly changing job market. This journey of self-improvement is not a choice but a necessity for those seeking to thrive in an ever-changing world.

Adults enhance their employability and build a resilient mindset by investing in their personal and professional development. They learn to navigate the dynamic employment landscape and adapt to evolving market needs. By doing this, they can stay ahead of the competition and protect their position in a tough economy. The impact of continuous learning extends beyond the individual, as their enhanced skills contribute to improved economic performance and a higher quality of life for their communities. The rapid development of AI and the Fourth Industrial Revolution is changing the world. Continuous learning becomes vital for adults in their careers.

The rapid obsolescence of skills in today's economy underscores the importance of prizing lifelong learning and upskilling. Soft skills, such as adaptability and critical thinking, are essential for navigating this dynamic landscape. The job market today demands a dynamic skill set, and the ability to adapt is now a necessity.

Upskilling is no longer an option but a commitment that professionals must embrace to future-proof their careers. To succeed in this rapidly evolving environment, individuals must adopt a lifelong learning approach. They should seek new skills to remain relevant and thrive in a dynamic world.

As technology advances, so too must the workforce. Adults who commit to lifelong learning gain a competitive edge, ensuring they remain valuable assets in a market that favors agility and versatility. Those who cannot adapt will be left behind as their skill sets become outdated and less marketable.

Soft Skills in Demand

So, what are the most in-demand soft skills that employees should cultivate to maintain their employability? Effective communication remains at the top of the list. Articulating thoughts, listening, and tailoring communication styles to different audiences are what employees value.

Employers seek problem-solving and critical-thinking skills because these skills enable employees to approach challenges and make informed decisions. Collaboration and teamwork are also essential, fostering a cohesive and productive work environment. The most popular skills for maintaining employment in this future landscape revolve around adaptability and soft skills.

Employers seek individuals with strong critical thinking and problem-solving skills, capable of navigating the complex, ever-evolving challenges of the modern workplace. Employers also value emotional intelligence and practical communication skills, as they foster collaborative and innovative work environments. Despite advancements and efforts to close the skills gap, the problem persists. Employers continue to struggle to find candidates who are both practical and adaptable.

The educational landscape of 2035, as described, reflects a profound shift, driven by the integration of remote learning, AI, and VR. This paradigm aimed to bridge the gap between education and industry by cultivating essential soft skills alongside technical competencies. The envisioned future classrooms will materialize as active, versatile learning spaces that transcend old-style buildings. To grapple with intricate ideas through

Virtual mentors powered by artificial intelligence guide students through role-playing scenarios, fostering analytical skills, troubleshooting abilities, teamwork, and empathy. AI assistants are helping to decentralize educational oversight, a shift that new laws and the Department of Education have accelerated. It empowers state and local structures, tech providers, and industry associations. It will lead to diverse learning pathways. Tomorrow's students will use digital tools and become allies in the new learning world. While engaging with their devices, a historical theory aligns with their present—Bandura's social learning theory. It suggests that students are not just unengaged consumers of data.

Despite these advancements, a lingering skills disparity remained a critical concern. While technology facilitated the delivery of technical competencies, consistently cultivating relevant interpersonal skills across educational centers and avenues proved challenging in the absence of a federal standards body. A significant number of academic professionals are transitioning to corporate learning, bringing their expertise and critical thinking skills with them.

The core of this educational paradigm lies in its dual focus on technical competencies and the cultivation of essential soft skills. AI-driven avatar mentors guide students through complex simulations, reinforcing observational learning and providing feedback on critical thinking, problem-solving, collaboration, and emotional intelligence. This approach, while robust and grounded in foundational social learning theories, continues to face scrutiny for its efficacy in replicating real-world interpersonal dynamics.

6

The Classrooms and Schools of Tomorrow

The Classroom of the Future

T he internet and artificial intelligence power the classroom of the future. Students now have access to a wealth of knowledge at their fingertips, and AI has transformed how they interact with information and with one another. Bandura's Social learning theory suggests that people learn to exhibit behaviors by observing others and recognizing their behavior. The internet has reshaped the educational landscape across all grade levels. For early elementary students, interactive online platforms deliver foundational literacy and numeracy skills through gamified lessons and personalized feedback, allowing observation of peers' successes and the creation of adaptive learning paths.

In middle school, the internet provides access to virtual field trips and collaborative research projects, enabling students to witness and emulate effective digital citizenship and problem-solving strategies demonstrated by classmates, encouraging them to adopt these strategies in global exchanges. High school students use the internet for advanced research, online tutoring, and virtual lab simulations. They observe their peers using digital tools for scientific inquiry and critical analysis, which encourages them to adopt advanced learning behaviors.

The history of distance learning, which began with correspondence courses sent by postal mail, is evolving into online education. These early iterations, though limited in interactivity, laid the groundwork for a more accessible educational model. Radio and television further expand distance learning, offering broadcast lectures and educational programs to broader audiences.

The internet, however, represented a significant change, moving from one-to-many broadcasts to many-to-many networked interactions. It facilitated synchronous and asynchronous communication, enabling real-time discussions, collaborative document creation, and the establishment of virtual learning environments that mirror many aspects of a physical classroom.

The integration of artificial intelligence and specialized online platforms has enabled personalized learning and access to resources, regardless of location. The classroom of the future is not just preparing students for academic success, but also for their transition into adulthood and the professional world. The Asia Society Center for Global Education (ASCGE) emphasizes the importance of developing essential soft skills —such as leadership, collaboration, and communication —to enhance adaptability and success in adulthood. AI plays a pivotal role in honing these skills through interactive and immersive learning experiences.

AI platforms create simulated environments that mirror real-world scenarios, allowing students to step into leadership roles and collaborate with their peers. It enables students to develop emotional intelligence by receiving feedback on their interactions and reflecting on their self-awareness and self-management skills. Bandura's social learning theory has revolutionized educators' understanding of the learning process. His three core concepts reveal how people learn to exhibit behaviors by observing others' behaviors and witnessing the reinforcement of these behaviors over time in this futuristic classroom.

AI-powered systems that can record and replay lessons facilitate the first concept, allowing students to observe their peers and mentors, as well as others' behaviors. To address the second concept —the role of internal mental states —the personalized learning paths consider each student's unique

75

cognitive needs. The third concept, a crucial reminder, is that the future classroom is not only about acquiring knowledge but also applying it, as learning may not always lead to change.

This futuristic classroom, guided by Bandura's social learning theory, creates a unique and personalized learning journey for each student. Social learning theories emphasize the importance of observing and interacting with others, and AI-powered classrooms offer students opportunities to collaborate and learn from their peers.

People can learn new information and behaviors by observing other people's reinforcement (Bandura, 1986). The future classroom brings a new era of learning. Powered by the internet and artificial intelligence, students immerse themselves in a wealth of knowledge. AI has transformed the way they interact with information and each other, fostering a dynamic and engaging educational experience.

Development of Soft Skills

The future classroom is a bustling hub of technological innovation, with students at the forefront of this revolution. With the internet and AI as their allies, they navigate a new world of learning. As they interact with their devices, a theory from the past resonates with their present—Bandura's social learning theory. It suggests that students are not just passive recipients of information; they are also observers and imitators, absorbing and mirroring the behaviors they witness to achieve success in adulthood.

Bandura's theory of observational learning is clear as they witness their classmates interact with AI tutors, collaborating on virtual projects, and engaging with augmented reality. This new educational paradigm has disrupted traditional methods with a profound impact. By observing their peers and mentors through AI-recorded lessons, students learn from a variety of role models.

Students are actively learning in the classroom. They recognize that social interactions are just as meaningful as their studies. They observe and emulate their peers, seizing new knowledge and skills. The classroom of the future is not just about accessing information but also about developing soft skills.

In American classrooms of the future, educators' roles have evolved. Teachers have become facilitators and guides, navigating students through a landscape of AI and technological innovation. Today's students experience a classroom of the future, where AI and technology have disrupted traditional learning methods. These students, a diverse group with varying levels of emotional intelligence, are now active participants in their learning journey.

This new classroom environment goes beyond academic learning, as social learning theories come into play. Students observe and emulate their peers' behavior. It is a place where emotional intelligence, as defined by Goleman, thrives. As students engage in online discussions and projects, they develop emotional intelligence — the ability to identify and manage their own emotions and understand others' feelings. It is crucial for building strong relationships and navigating social situations.

2035 Educational Landscape

The classrooms of the future will need to bridge the gap between education and industry, addressing the soft skills deficit among graduates. By 2035, integrating remote learning, artificial intelligence, and virtual reality will be pivotal in equipping students with industry-relevant competencies. Schools will embrace a hybrid model, blending physical and virtual classrooms. Immersive VR experiences will deliver lessons, enabling students to explore complex scientific concepts and integrate technology through interactive simulations. AI will play a pivotal role, with intelligent tutoring systems providing personalized learning experiences tailored to each student's unique needs.

Integrating remote learning, artificial intelligence, and virtual reality holds promise for both education and industry. However, it adds logistical complexity. The hybrid model, which combines physical and virtual classrooms, necessitates robust systems to track student engagement across both modalities, particularly when processing financial transactions. Intelligent tutoring systems, designed for personalized learning experiences, should also be integrated with financial aid and payment processing databases to prevent inadvertent drops because of systemic delays.

AI-driven avatar mentors help students. Their goal is to guide students through interactive role-playing scenarios. It teaches critical thinking, problem-solving, collaboration, and emotional intelligence. However, financial issues limited their effectiveness when students stopped participating.

The burgeoning enrollment figures of 2035 will stand in stark contrast to the operational realities faced by many educational institutions. A systematic review of student records will reveal that a significant number of learners drop their courses due to outstanding payments. This administrative action, referred to as a "purge," will continue, with an impending attendance verification process set to identify and drop further students.

This situation resulted from several converging systemic changes. These included the new verification requirement for all students and a rise in students paying for their education. These factors have increased the number of class drops and will further escalate demand for student reinstatement procedures. The administrative burden generated by these reinstatement requests will become a focal point for institutional analysis.

Institutions face a dual imperative. They must advance pedagogical reforms while also addressing student retention and financial challenges. The successful implementation of virtual reality simulations and AI-powered soft skills development hinged on a stable and enrolled student body. Therefore, a critical review of payment processing workflows and communication protocols related to financial obligations became a necessary precursor to realizing the transformative potential of the 2035 educational landscape.

Integrating remote learning will extend the reach of these advanced classrooms beyond geographical limitations. Beyond technical skills, educational environments of 2035 will prioritize fostering emotional intelligence. Virtual mentors powered by AI will guide students through engaging simulations, helping them cultivate essential soft skills such as analytical reasoning, troubleshooting, teamwork, and empathy. The forthcoming education model will ensure that graduates are better prepared to meet the requirements of a rapidly changing job market, thereby bridging the divide between academia and industry.

Students unable to attend physical sessions will participate in immersive VR experiences, engaging with AI-driven avatars and peer groups, guaranteeing fair access to industry-relevant competencies. AI will analyze communication patterns, collaborative contributions, and problem-solving approaches within these interactive scenarios to assess soft skills. This holistic education prepares graduates for the future. It focuses on technical skills and interpersonal abilities.

The physical spaces of these future schools will transform from old-fashioned buildings into dynamic, modular environments. Instead of fixed rows of desks, learning hub systems allow students to access global supply chains. Specialized AI-assisted labs will provide access to advanced fabrication tools and quantum computing simulators, all overseen by AI diagnostics that ensure safety and optimize experimental parameters.

These educational institutions will further distinguish themselves through their flexible scheduling and learning pathways. Students will progress through competency-based modules, rather than adhering to rigid grade levels, allowing for accelerated mastery of specific skills and deeper exploration of others.

A single, full-time teacher will lead the program. Industry professionals will help develop the curriculum. They will also give guest lectures online or in advanced learning environments. It strengthens the connection between education and practical use.

The Dispersal of Brick-and-Mortar Campuses

The fourth industrial revolution will lead to a significant transformation in the physical footprint of educational institutions. The shift towards dynamic, modular learning hubs, rather than static, traditional brick-and-mortar structures, will reduce the need for extensive, single-purpose buildings. Adaptable spaces are essential for reconfigurable furniture, interactive walls, and dedicated VR immersion zones, enabling people to repurpose them with relative ease. This modularity, combined with the seamless integration of remote learning, suggests a decentralization of physical campuses.

Instead of large, centralized schools, cities might see the proliferation of smaller, flexible learning hubs within communities, or even repurposed existing commercial spaces. This reimagining of school real estate could reduce the overall number of traditional school buildings required, freeing up valuable urban land for other uses. The impact on urban real estate will be substantial. The decline in demand for large, conventional school buildings could lead to repurposing opportunities.

Developers could convert existing structures into mixed-use developments, community centers, or affordable housing, thereby addressing other urban needs. Educational facilities could respond to shifts more easily by scaling up or down. It is due to the modular nature of future learning environments, which rely more on portable infrastructure. This physical space flexibility could shape distributed learning networks. A central hub might support smaller learning nodes, diversifying the city.

The real estate market will see a recalibration of educational property values. The demand for traditional school campuses may decrease. There will be increased demand for adaptable spaces for modern learning. These renovated or built educational facilities. This evolution alters the relationship between education and urban planning, influencing how cities allocate and use their land resources.

People should expect a decline in the demand for large, single-purpose, traditional brick-and-mortar structures as schools develop into flexible, adaptable learning centers. This shift signifies a potential decrease in the total amount of conventional school buildings required within cities—the dispersal of brick-and-mortar campuses. Instead of sprawling, centralized institutions, cities may witness a spread of smaller, distributed learning nodes. Placing these hubs in various urban environments, including renovated existing commercial spaces, could reduce reliance on extensive, purpose-built educational facilities. This reallocation of land resources could free up significant urban acreage for alternative development, addressing other community needs such as affordable housing or green spaces.

Adaptive Reuse

The concept of adaptive reuse will become central to the educational real estate sector. Existing, underutilized structures, ranging from former retail centers to adaptable industrial buildings, will be prime candidates for conversion into modern learning environments. The spaces require flexible furniture, interactive walls, and VR environments. Repurposing existing buildings is a sustainable and economical approach that reduces the environmental impact of new construction.

The modularity of future learning environments allows for scalability. Educational facilities can adapt more easily to changes. It encompasses demographic shifts and innovative teaching methods, leading to a more adaptable educational system in cities. The recalibration of educational property values is an inevitable consequence of these changes. The demand for significant, traditional school campuses may decrease. However, adaptable commercial or industrial spaces redeveloped into learning hubs are likely to see increased demand.

Investment in sophisticated technological infrastructure, such as haptic feedback suits, full-body tracking systems, and AI-assisted labs, will dictate the specific requirements for these repurposed or newly constructed educational spaces. This fundamental alteration in school design and functionality will cause a re-evaluation of urban planning strategies. Cities must proactively work with schools. It will help them use and distribute land. The goal is to use educational progress to benefit cities and communities.

Teacher Shortages

The wave of teacher retirements looming ahead presents a significant challenge to the successful implementation of these future educational models. Researchers predict that a substantial portion of the current teaching workforce will retire, creating a critical shortage of experienced educators. Younger generations show little interest in teaching. The perception of demanding workloads and low pay exacerbates the situation.

AI-driven tutoring systems and virtual reality simulations are giving rise to a new type of educator. This educator must be skilled at teaching in complex digital settings and helping students build essential soft skills. The current pedagogical pipeline struggles to produce enough individuals equipped with both the technical acumen and the interpersonal facilitation skills necessary to thrive in these developed roles.

To address the upcoming teacher shortage, we must recruit new teachers and support current ones in their professional development. The evolving role of the educator, shifting from lecturer to facilitator and mentor, necessitates a reevaluation of teacher training programs. These programs must incorporate advanced digital literacy, pedagogical approaches for immersive learning environments, and the effective integration of AI-powered tools.

Initiatives to elevate the status and attractiveness of the teaching profession are paramount. It might include better pay, opportunities to learn about the latest teaching technologies, and groups where teachers can share ideas and collaborate on lessons. Industry professionals, as mentioned, will play a vital role in bridging this gap by contributing expertise and mentoring aspiring educators.

The success of these advanced educational institutions by 2035 depends on the effectiveness of their education systems. They must either develop new teachers or keep current ones to meet the needs of these changing learning environments. Integrating remote learning, AI, and virtual reality while promising to equip students with industry-relevant competencies will falter without a robust, adaptable teaching workforce. Balancing technology with educators' central role in evolving classrooms remains a challenge.

Inherent Skills to Innovate, Adapt, and Lead

AI plays a pivotal role in this future education model, with intelligent tutoring systems tailoring lessons to each student's unique needs. These systems adapt to learning styles and paces, providing personalized guidance and ensuring a deep understanding of the material. Nevertheless, the classrooms of 2035 go beyond imparting technical knowledge. Soft skills, once considered a deficit among graduates, are now at the forefront of the curriculum.

AI-driven avatar mentors, with their infinite patience and customized approaches, guide students through a myriad of interactive role-playing scenarios. These scenarios challenge students to apply critical thinking and problem-solving skills, encouraging creativity and innovation. They learn the art of collaboration, working together to tackle complex projects, and developing emotional intelligence as they navigate diverse perspectives and virtual team dynamics. These learners are not just absorbing information; they are constructing knowledge, and their soft skills are developing through real-world applications within simulated environments.

The AI-driven avatars not only present scenarios; they prompt critical thinking by posing challenging questions, encouraging students to justify their decisions and consider alternative perspectives. For some, this means Mentor nudging them to analyze the historical context of their trade negotiations and to assess the long-term societal effects of their proposed agreements. For others, it involves the AI tutor posing hypothetical engineering problems that require them to apply their calculus knowledge, pushing them to think beyond rote memorization and toward creative problem-solving.

The classroom of the future is here, powered by the internet and artificial intelligence. Technology and humanistic guidance are merging. It produces graduates who excel, are adaptable, intelligent, and prepared for today's workforce. Students now have access to a wealth of knowledge at their fingertips, and AI has transformed how they interact with information and with one another.

The ripple effect of this educational revolution is palpable. As students emerge from these sophisticated learning environments, they bring a unique blend of technical mastery and honed interpersonal skills. The demand for graduates who can transition from intricate simulations to real-world collaborative projects is soaring. This success validates the initial vision: classrooms that bridge the gap between academia and industry, producing individuals equipped for the future —not just with knowledge but with the inherent skills to innovate, adapt, and lead.

The Future of the Department of Education

The Federal Department of Education, established in 1979, is expected to cease operations soon. Some Republicans have been calling for the department's closure since its establishment. In 2024, 24.64% of Republicans and Republican-leaning Americans had a negative view of the Department of Education. The future of the Department of Education symbolizes an ideological divide over the extent to which the federal government should regulate education.

Abolishing the Department of Education appears to be only one aspect of the broader reforms outlined in Project 2025, a collection of policy plans developed by conservatives. Project 2025 frames a discussion of significant changes in federal education policy around the idea that American schools are underperforming.

Project 2025 also proposes block grants of federal funds for low-income students, allocated through Title I. Under the plan, the block grants would be available for several years, after which the federal role would phase out entirely, leaving the states to decide whether to provide that funding.

Hypothetically, let us assume that by 2034, following the passage of a new Education Act, oversight and funding will be decentralized, shifting authority to state and local governance structures, as well as to a consortium of educational technology providers and industry guilds. Closing the department would also involve transferring archival data to the National Archives, ending grant programs for traditional education, and reassigning staff to agencies focused on education standards and workforce development.

Consolidating many administrative functions at the state level will be a direct consequence, requiring states to develop new regulatory frameworks and funding mechanisms. Developing industry-relevant competencies, once a directive from federal policy, will become a direct outcome of collaborative efforts between educational providers and the guilds themselves. This decentralization will have profound spatial implications. As mentioned, the shift towards modular learning hubs, often repurposed commercial or industrial spaces, will reduce the demand for traditional, large-scale school buildings.

The immediate impact of the Department of Education's closure would be a significant recalibration of the educational landscape. States would be responsible for defining curricula and accreditation standards, potentially leading to a surge in localized educational models. Funding streams would shift from federal grants to a combination of state-level taxation, private sector partnerships, and direct tuition models facilitated by accredited technology providers.

The consortium of education guilds, authorized by the recently enacted legislation, would confer industry-specific certifications that often took precedence over conventional university degrees as qualifications for entry-level roles. This recalibration will lead to diversifying educational pathways, with competency-based modules and skill-specific bootcamps gaining prominence alongside the remaining brick-and-mortar institutions.

The closure of the federal department will accelerate this trend by eliminating a layer of oversight that occasionally dictates standardized building requirements. States and local districts, while freed from these federal mandates, could more readily approve the conversion of existing structures into adaptable learning environments. This urban real estate flexibility allows us to place smaller learning nodes in communities. It reduces long commutes and integrates education more effectively into cities. Demand will continue to shift toward adaptable spaces with robust technological infrastructure, thereby reinforcing the new paradigm.

Modern Educational Paradigm

Despite these advancements and the optimistic outlook projected for 2035, a persistent skills gap remains a critical concern for both public and private educational institutions. Schools face challenges in preparing graduates to meet industry demands, despite the use of advanced technology. Patacsil and Tablatin (2017) found that many graduates lack sufficient soft skills for the job market, a key factor in the skills gap. It is true in science and technology (Ahmad et al., 2019).

While AI-driven avatar mentors and immersive simulations aim to cultivate these competencies, the efficacy of these tools in mitigating the deficit remains under scrutiny. The decentralized educational system, while fostering flexibility, has also posed new challenges in ensuring the uniform development of soft skills across diverse learning hubs and pathways. While graduates may demonstrate proficiency in specific technical competencies, their ability to collaborate, communicate complex ideas, and engage in critical problem-solving within real-world, interdisciplinary contexts may still fall short of employer expectations. It leads to the ongoing refinement of AI feedback systems and virtual-character training environments to better reflect the nuances and complexities of actual professional interactions.

In this modern educational paradigm, AI avatars and virtual human facilitators might not replicate spontaneous interactions. This lack of experiential learning poses a significant challenge to current educational models, particularly in developing soft skills. The academic environment in the coming years will be a lively center of activity, with students immersed in interactive lessons.

The space is abuzz with the gentle hum of advanced technology, as intelligent systems facilitate a unique learning experience tailored to each student. It is a far cry from the traditional classrooms of the past, where learning was a one-size-fits-all affair. Now, the insights of pioneers like Bandura have forever transformed the educational landscape.

As the students interact with their lessons, they focus on more than just the information on their screens or holographic displays. They are attentive to their peers and the behavior shown around them. They witness their classmates' successes and failures and, using Bandura's framework, realize that learning is not an individual pursuit but a social one. This classroom mirrors the wider world, emphasizing the significance of teamwork and attentive watching as much as independent learning.

The decentralized educational system further compounds this ongoing challenge. This shift promotes flexibility and the development of modular learning hubs. However, it also creates challenges in ensuring consistent soft skill development across different learning environments and pathways.

The absence of a federal body to establish overarching standards has led to divergence in curriculum design and the assessment of these crucial non-technical abilities.

Graduates may possess strong technical skills but often lack the soft skills that employers seek. Collaboration, communication, emotional intelligence, and problem-solving abilities may be lacking. It causes the continuous refinement of artificial intelligence-powered feedback systems and virtual character-led training simulations. Virtual environments with AI avatars or human facilitators may not simulate natural interactions. This deficiency tests current educational models.

Availability of Educational Professionals

The sustained efficacy of these futuristic classrooms and the bridging of the skills gap also depend on the availability and adaptability of educational professionals. As the demand for industry-relevant competencies soared and pedagogical models shifted toward facilitation and mentorship within immersive technological environments, a significant migration of academic professionals began.

Educators employed in traditional university settings, as well as some K-12 districts, have found their expertise sought after by Learning & Development (L&D) departments within corporations and specialized educational technology providers. These training jobs provided more experience with the newest technologies, often with better pay, and a better chance to put into practice the educational changes planned for 2035. Schools lost experienced educators as they moved away, creating a challenge for developing learning hubs. At the same time, the schools themselves struggled, needing to change or face closure.

This demographic shift in the workforce meant that, while the theoretical framework for future classrooms was robust, the practical implementation faced a critical shortage of human capital. Industries are now recruiting individuals who have supported remote learning, AI, and VR in academia. The industries want to use their knowledge to train and upskill. The institutions' success in bridging the soft skills gap depended on their ability to attract and keep professionals. These professionals needed to understand observational learning, reinforcement, and the application of knowledge in real-world contexts.

The consortium of educational technology providers and industry guilds, envisioned as collaborators, became primary employers, generating educators adept at navigating the new digital learning paradigms. This dynamic also restructured how they sourced and developed educational professionals. It emphasized rapid upskilling programs and created flexible, project-based roles that mirrored the modular nature of the learning environment.

Therefore, people can see the persistent skills gap not only as a curriculum or technological deficiency but also as a symptom of this widespread migration of talent. As academic professionals transition into L&D roles, the intellectual capital and practical experience needed to refine AI-driven feedback mechanisms and avatar-led training scenarios within schools are dispersed across other areas. Due to the challenging job market, schools are required to adjust their hiring and retention strategies. They began offering incentives, including bonuses and training, to attract and retain employees, especially since other jobs offered higher pay.

Beyond the Visual

Classrooms will be significantly different by 2035, utilizing innovative technology and new teaching methods to make learning more engaging and enjoyable. Designers will create the physical space with flexibility in mind, including modular furniture and interactive walls that can adapt to different lessons and activities. Virtual reality and augmented reality will play a significant role, allowing students to explore complex concepts and visit far-flung places without leaving their seats. History lessons could utilize VR headsets, allowing students to experience historical events firsthand. Science classes could use virtual labs for experiments that incorporate digital tools and simulations.

Beyond the visual, the future classroom will be a multisensory experience. Soundscapes and haptic feedback will enhance the learning experience. For example, students could feel the vibrations of a dinosaur stomp during a prehistoric life lesson. They might smell baking bread in a culinary arts class. This multisensory approach will engage students at multiple levels, catering to different learning styles and creating a more immersive, memorable educational journey.

The classroom will integrate artificial intelligence. AI tutors will provide personalized learning experiences, adapting to each student's unique needs and progress. These intelligent systems will offer tailored lessons, instant feedback, and targeted interventions to ensure every student receives the support and challenges they need to thrive. Natural language processing will enable students to interact with their learning environment through voice commands and conversational interfaces, making technology an intuitive and accessible tool for learning.

The future of education is an exciting prospect, with technology enhancing and transforming the way students learn. By embracing these innovations, educators can create a dynamic, immersive learning environment that captivates their students. The classroom of 2035 is a place where students are engaged and enthralled, their senses stimulated, and their curiosity piqued.

Students experience other worlds with haptic feedback and soundscapes. In their classroom, they study aerodynamics of flight and experience the feeling of wind. They also sense the echoes of historical battles, as if the clashing swords are still resounding.

The applications of virtual and augmented reality are boundless, allowing students to explore the human body, travel through space, and interact with complex data visualizations. Artificial intelligence is the quiet companion, always ready to offer a helping hand. AI tutors provide personalized learning paths, tailoring to each student's unique needs and ensuring every student achieves success. Natural language processing enables seamless interaction with technology, resulting in an intuitive and accessible learning experience.

Schools of Tomorrow

In conclusion, future classrooms represent a significant evolution in educational delivery, integrating advanced technologies to foster both technical proficiency and the cultivation of essential interpersonal skills. It is challenging to align academic preparation with industry readiness perfectly. This persistent challenge requires innovation in teaching and investment in educational staff.

The success of these advanced educational institutions by 2035 hinges on the continuous adaptation of pedagogical approaches and the strategic deployment of human capital. Migrating experienced educators to corporate learning and development (L&D) roles highlights a critical bottleneck, necessitating innovative recruitment and retention strategies within academia.

As the world moves forward, the classrooms of 2035 stand as a testament to a bold vision of education —one that is sophisticated, interdisciplinary, and responsive to industry's demands. Soft skills, AI-driven learning environments, and educators are key. This combination will shape the future of learning, where graduates experience constant change and the relationship between humans and AI.

References

Ahmad, S., Othman, S., Ismail, M., & Abd Rahman, M. (2019). The relationship between soft skills and the employability of STEM graduates. *International Journal of Academic Research in Business and Social Sciences*, *9*(11), 1043-1057.

Callea V, Remigi G, Tempone R, Matrisciano L, Ursache M. COACHING ING EDUCATIONAL MODEL: analysis and application for business. IEEE Global Engineering Education Conference, EDUCON. 2022; pp. 445-450.

Carvalho, Carlos A. The Adequacy of Accounting Education in the Development of Transversal Skills Needed to Meet Market Demands. Sustainability. 2022;14(10):5755.

Christensen EL, Paasivaara M. Learning Soft Skills through Distributed Software. 2022.

Fouzia M. More than technical experts: Engineering professionals' perspectives on the role of soft skills in their practice, Industry and Higher Education. 2022;36(3):294–305.

Gascóna ÁE, Gallifab J. How to measure soft skills in the educational context: psychometric properties of the SKILLS-in-ONE questionnaire. Studies in Educational Evaluation; 2022.

Google Oxygen Project: Do Managers Matter? Harvard Business School. 2020; pp. 223.

Google Project Aristotle [Internet]Available from: https://www.inc.com/michael-schneider/google-thought1023they-knew-how-to-create-the-perfect.html. [Accessed 10 August 2022].

Horváth G, Juhász T. A puha és a kemény készségek munkaerőpiaci szükségessége, Education. 2021;30(3):532–42.

Horváth G, Juhász T. The Emergence of Soft Skills in Agricultural Education.

Problems and Perspectives in Management 2021;19(3):453-66.

Knowles, M. S. (1984). *Andragogy in Action: Applying Modern Principles of Adult Education*. Jossey-Bass. Mezirow, J. (2000). *Learning as Transformation: Critical Perspectives on a Theory of Adult Learning*. Jossey-Bass. shown

Kolb, D. A. (1984). *Experiential Learning: Experience as the Source of Learning and Development*. Prentice-Hall.

Kyrousi AG-E, Leivadi S. Business employability for late millennials: exploring the perceptions of generation Z students and generation X faculty. Management Research Review. 2022;45(5):664-83.

Kárpátiné DJA kompetencia hatalom? Mitér a gazdasági felsőfokúképzés, ha nem ad diplomat? 2014. Available from: http://realphd.mtak.hu/624/1/Kar patineDarocziJuditdissertation.pdf.

Laker DR, Powell JL. The Differences between Hard and Soft Skills and Their Relative Impact on Training Transfer. Human Resource Development Quarterly. 2011;22(1)111-22.

Lepeley M-TSoft skills: The lingua franca of human-centered management in the global VUCA environment. In Lepeley M-T, Beutell NJ, Abarca N, Majluf N, editors. Soft Skills for Human Centered Management and Global Sustainability. 2021; pp. 3-22.

LinkedIn: Data-driven insights into the changing world of work [Internet]. Available from: https://business.linkedin.com/talent-solutions/resources/tal ent-strategy/global-talent-trends-2019. [Accessed 10 August 2022].

McLaren Candidates' Soft Skills are Notoriously Hard to Assess, but Following These 6 Steps Will Help. Linkedin. 2019;Available from:https://www.linke din.com/business/talent/blog/talentacquisition/soft-skills-are-hardtoasses s-but-these-steps-can-help.

Miller JJM, Zakary PLA, qualitative analysis of undergraduate sport management student skill and awareness development at an international sports event, Journal of Hospitality, Leisure, Sport and Tourism Education. 2022;30.

National Soft Skills Association: The Soft Skills Disconnect. 2015. Available from:https://www.nationalsoftskills.org/the-soft-skillsdisconnect/.

Patacsil, J. V., & Tablatin, R. A. (2017). Bridging the skills gap: Does the higher education curriculum meet the industry's needs? *Proceedings of the 2017 International Conference on Management, Education and Social Sciences*, 278-283.

Rajabzadeh A, Long J, Zeadin M. Engineering Student Experiences of Group Work. Education Sciences. 2022;12(5).

Rasli M, Ghani FA, Razali NHA, Razak SFFA, Razak MZA, Embong F, Salleh NSA, Idris RSNR, Rani SM. Do Soft Skills Really Matter? In Othman NS, Jaaffar AHB, Harun NHB, Buniamin SB, Mohamad NEAB, Ali IBM, et al., editors. Driving Sustainability through Business-Technology Synergy, vol. 100. European Proceedings of Social and Behavioral Sciences (pp. 427–435).

Knowles, M. S. (1984). *The Adult Learner: A Neglected Species*. Gulf Publishing Company. Kolb, D. A. (1984). *Experiential Learning: Experience as the Source of Learning and Development*. Prentice-Hall.

Robles MM Executive Perceptions of the Top 10 Soft Skills Needed in Today's Workplace. Business Communication Quarterly. 2012;75(4):453–65.

Scheerens J, van der Werf G, Boer H, editors. Measurement of Soft Skills in Education. Soft Skills in Education. 2020. pp. 141-189.

Schulz B. The Importance of Soft Skills: Education Beyond Academic Knowledge. NAWA Journal of Language and Communication. 2008;2(1):146–54.

Seetha S. Necessity of Soft Skills Training for Students and Professionals. International Association of Scientific Innovation and Research. 2013;4(2):171–4.

SEMCOG. (2012). *Workforce Development System Strategy*. Southeast Michigan Council of Governments.

Sopa A, Asbari M, Purwanto A, Santoso PB, Mustofa, Hutagalung D, Maesaroh S, Ramdan M, Primahendra R. Hard Skills versus Soft Skills: Which are More Important for Indonesian Employees' Innovation Capability, International Journal of Control and Automation 2020; 13(2), 156–75.

The Wall Street Journal: Employers Find 'Soft Skills' Like Critical Thinking in Short Supply, 2016. Available from: https://www.wsj.com/articles/employers-find-soft-skillslike-critical-thinking-in-short-supply-1472549400.

Tribble LSThe importance of soft skills in the workplace as perceived by community college instructors and industries (Doctoral Thesis). Mississippi

State University. 2020. Available from: https://ir.library.msstate.edu/bitstrea m/handle/11668/17062/etd-11022009-172940.pdf.

Tripathy M Relevance of Soft Skills in Career Success. MIER Journal of Educational Studies Trends & Practices. 2021;10(1):91–102.

Tulgan B. Bridging the soft skills gap: How to teach the soft skills basics to Today's Young Talent. Wiley; 2015.

World Economic Forum. (2020). *The Future of Jobs Report 2020*.

About the Author

Dr. Paul is a trusted business and technology consultant for the world's top investment banks, private equity firms, Fortune 100 companies, Real estate firms, and business schools. Dr. Paul develops adaptive learning programs and helps businesses scale by leveraging AI. He's partnered with The Second City to develop improv workshops and written Keynote speeches for some of TED Talks' most prominent speakers. He also has extensive experience in quantum computing, augmented intelligence, training, teaching, coaching, and leadership development.

Dr. Paul specializes in the rapid development of end-to-end eLearning, leadership programs, and Skill Labs. He is a skilled cyberpsychologist and methodologist. He has worked with more than a hundred Learning Management Systems. He leverages his knowledge and experience, as well as Artificial Intelligence. Dr. Paul is skilled in developing and implementing the full learning cycle from Analysis to Evaluation. He helps elaborate a balanced feature set that caters to diverse learning needs and maximizes the value of LMS adoption.

Dr. Paul is a member of several associations. He earned a bachelor's degree in business, a master's in project management, an M.B.A., and an Ed.D.. He is known for being passionate, reliable, and a producer of quality. As an advocate for entrepreneurs and real estate investors, he's always had an intrinsic motivation to help others achieve their goals.